Knowledge Management:

An Optimization Challenge

By Leonardo Mora

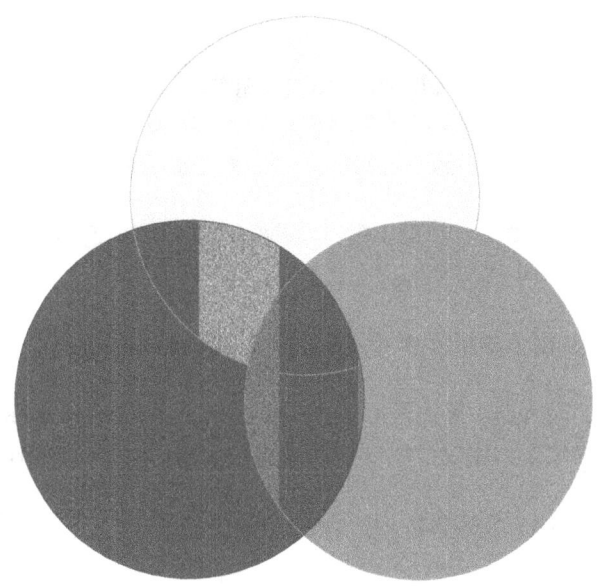

ISBN 978-0-6151-5390-2 | Printed in the USA | First Edition.

Acknowledgements

For all the patience and hard work, to my father Julian, brother Luis who helped me out figuring out how to write this book.

I give thanks to those who contributed in the website with great information and opinion.

Table of Contents

Prologue

To my wife Lucia, Isabella and Sebastian.

"Any sufficiently advanced technology is indistinguishable from magic. " wrote the British scientist and author Arthur C. Clarke *"Profiles of The Future", 1961 (Clarke's third law),* in other words when the human being does not understand advanced technology, it considers it magic. And magic can produce positive and negative reactions, the first called curiosity and the second rejection to the unknown. History teaches us that when the personal computer[1] was born, the reaction has been the latter, which explains the delay we've been having understanding what the possibilities are using PC's as a communications tool to enable collaboration.

The Human Factor (Collaboration or Cooperation[2]) drives the creation, development, implementation and enhancement of science and technology in general..
In the majority of countries, the official government resists and rejects letting it's knowledge be known to the people. Same principle applies to private companies. The controlling nature and autocracy structure generates silos — Organizational as informational. Human condition copied from nature (Preservation or survival of the strongest) prescribes each individual must defend fiercely its survival conserving any advantage physical or intellectual that differentiates him from others; sharing them takes

---

[1] We will understand "Personal Computer" to refer to an affordable, general-purpose, microprocessor- based computer intended for the consumer market.
[2] Collaboration definition: To work together, especially in a joint intellectual effort.: **Cooperation**: How groups of people can form networks of trust without a central system controlling their behavior or directly enforcing their compliance. James Surowiecki

away the character that makes him (her) possess a privileged spot in society.

In the beginning of the information age alter 1950, the popular question between people was "How many people will the computer replace because it can make thousands of operations per second? In the mid 1980's that question evolved into "How can I use the personal computer?"

Once into the information age distributed in the Global Village through The InterNET, the human factor comes again as the main player (Demonstrating a big shift) with fresh evidence that information availability through services like Google, Wikipedia free to all, forces individuals to rethink what is the key to innovation and creativity, the next phase once you make all information available and accessible to anyone. The paradigm "Information is power" has changed into "The power of Collaboration".

In the corporate environment the fight is even harder given people's competitiveness. Each member might not be willing to deposit what they know[3] in the central "warehouse" until he (she) understand through better (Transformational) leadership that team learning and collaboration are the key to a successful (i.e, creative, innovative) organization.

Knowledge Management practitioners has then to rethink the way to approach individuals to make them share/collaborate their knowledge into one pool and give the right tools to make better and simple use of it. The objective might be reached when people start thinking that technology is not magic, and begin using it for its own benefit, learning and growth.

By being open minded to the ideas in this book you can realize the potential we've lost and the extensive knowledge possessed by humanity is ready for us to create a feasible solution where we can simplify the tools.

----

[3] Know as information acquired and the past lessons from experiences lived.

Why this book?

I've come to realize that the "Human Factor" and other issues invalidate Knowledge Management (KM) (an information technology area) effectiveness through my experience in multiple projects.

My first conference about KM (15 participants) and what is wrong with it became a reality in Poughkeepsie, NY in 2005. One of the participants was a scientist which wanted to expand these insights further and by looking at his fascination at me talking about the topic is what contributed to start this book .The content of it is a summary of 10 years worth of experiences, failures, hard lessons and learning about KM projects implementation in information technology; how we tried to apply it and the tough experience learning through trial and error.. I am delighted to share this e-book and invite you to collaborate to it with your own insights and experience on the website www.kpreview.com. Nothing new, just rediscovering the basics. And, should I be so presumptuous as to believe I can add to what other great minds have wrought? with humility I shall go on.

About the author.

In the beginning…

I was born and raised in Colombia. I learned English in high school and, in part, my father pushed me to have a British accent which I never caught quite right. I graduated in 1996 as a Systems Engineer from the Colombian School of Engineering, a high quality university down in Bogotá. I now realize I was not fully aware what the career name meant or encompassed at the time.
I had a quick stint at a research lab finishing my thesis project when I got an offer to work in a giant fast food US company as

their systems analyst. It went well until I finished school and they decided fulfill their promises when they hired me. Soon after that, I left in protest not before making a study to show they needed three full time people for the IT department, not one part time freshman. It was my first experience learning about what the corporate world was about. If it is not in writing you can pretend it was never said. After doubting about the decision I felt it was the right one. That first job was in essence do-it-all kind of thing. I managed everything in Information Technology (IT).; I always had some taste for having a broad interest for technology and I was been given the chance to put it into practice. I did not know how important that would be later on.

Soon after, I was called for an opportunity to do some software training courses as a consultant to an oil company -- a big one that is. The project: A huge migration from Apple to PC and we were the ones training ex-pats (i.e., Foreign people transferred to a location). I was about to witness my first large full scale IT project. That would not be very unique if I didn't tell you the migration was for two thousand plus people in less than three months. That is an enormous undertaking for any size of company.

They (training consultants) trained us how to train users, and it definitely helped me tune up my presentation skills plus a little of British accent refinement... In summary, it was a hard and eye opening experience. At the end, the project was a tremendous success. Soon after I was hired and transferred to work as a support analyst and part time Webmaster. The Internet was all the rage4, and the top job you could think about on an Internet environment was being a Webmaster. What I did not know was how hard it was to become one. The person doing the Webmaster "juggling" left to Canada,. After a while I finally got the job, but I was starting to get interested in the ERP (e.g., SAP) arena where salaries where all the rage and the technology looked great (So I thought).. It was that good.

Meanwhile I had to deal with this tool for document management that looked pretty interesting but that I did not see much of a future and potential in the webmaster area (How wrong I was). The tool was called Livelink. In the late 1990's I went from webmaster into being the tech lead guy for a big time project implementing

---

[4] Go to appendix 1

"Knowledge management" technology (the Livelink product I mentioned) to the entire organization. At that time, KM was unknown, and many initiatives where exploring how to do it.. Figuring out who was to be the project manager posed a challenge at the time, so managers decided that the best option would be to select a experienced operations manager from the field as the PM for this knowledge project. It did not take long to find out this would require a lot more than field experience. We did not know how big and complicated the project could be as well. Many observers and colleagues warned me left and right about this project and its dangers, but I felt compelled to keep trying anyway. In summary, the project had three different project managers (definitely not recommended on any kind of project), and we were lucky to have a system in place at the end with the best technology around cause we were able to bypass the many regular slow processes. The only person working full time in the ... continues Appendix 1

Why the human factor invalidates KM?

For those of you out of the technology field, the KM term stands for Knowledge Management, a buzz word created in the late nineties to imply a practice within IT to implement a set of software tools with the goal of centralizing and controlling information, documents, websites, etc in order to share them between the "right" people in a given organization. By right people I refer to a given permission structure behind all documents published. The premise is that those permissions would give access to a reduced group of people or individuals. This way you would "deliver the right information and documents to the right people. The KM goal originally was to centralize information and have people store information in a series of containers, websites, folders, etc expecting users to come back and make a natural use of it, letting you *find* the correct piece of information at the right time, imposing strict rules on who got to see what. It meant in reality trying to "*force*" people into sharing their knowledge through a set of tools and policies. Even though we were able to implement the technology successfully, I found that people sharing documents and in general their knowledge not to be the case, especially in the long run. Once we implemented the technology, people would randomly use it, never sharing their entire set of knowledge, or refusing to do it at all, and in some cases where policies where non existent, the use of a given system would die or come to a complete stop overtime. I found it is impossible to convince people to share their knowledge, if the culture of force prevails.

I also realized there is a complexity buildup (Managing the many different sources of information and groups) in the same autocratic structure which made it difficult to maintain or even scale in the long term. In other words, bureaucracy structures are already complex, adding permissions and twenty or thirty additional modules for all kinds of functionality to documents and information proved to be maddening.
The first symptom I witnessed was the enormous effort required to bring people into actively use the knowledge system. People

usually did not respond easily to a management request for doing so, unless managers themselves used it first. It happened in one project, managers expecting people to obey, but they would not use the software themselves.

Persuasion proved to be good in the short term in some cases through prices and rewards, but I noticed people would reluctantly use or take advantage of it. There was no clear goal as a whole as to why they had to put their information in a shared space, if their C drive was working fine. . There was a tremendous focus on the technical side (The general mentality of "if x.y.c company uses it", it should work for us), but the same or more effort was not invested into the people who would actually produce the goals you were looking for; the quality of training, coaching, user expectations, etc.

When we implemented digital processes (in IT language *Workflows*), the first project I did seemed to be successful; people used actively the system, but a series of mishaps ended up derailing the effort. The software vendor decided it was a "good idea" to change how people would interact with the system in a given process. That decision from an external agent crushed the internal effort in the company. Again, there was no clear strategy as to how it should be done avoiding this type of risk. We relied only on what the vendor software could do and no more.

After some reflection, I came to the conclusion that in order to create an environment where people feel comfortable and motivated to contribute, brainstorm ideas and interact freely, we needed a different approach; less restrictions , opening to a new way of thinking and leading people into collaboration because it was clearly not working. Focusing on technology only was not the path to success, so I learned.

The Path To Enlightment

Some of the initiatives I worked with in the past proved to be successful but they did not cover the wide scope usually companies engage upon deciding to go into knowledge management[5].

After joining NYU for a Project management course, in one of the classes –Risk Management-- we had to choose a book and make a presentation with what we learned. I probably chose the most difficult of all, but certainly the one I needed to find an answer to my problem.
In P.Senge's book (The Fifth Discipline) it reads "Tapping people's life long learning willingness and commitment" as a way for you to inspire genuine collaboration and cooperation. P. Senge explains that in order to do it, we need to start unlearning many of the wrong beliefs about how we work in the typical autocratic environment, and learn about systemic thinking and its disciplines. I realize is not a theory wildly put into action today, because it requires **openness**, and making a wild guess, it is something many leaders are not prepared or willing to do now.

Another aspect of knowledge management implementations hurting the overall outcome was people's needs, or the lack of attention to them. Top management had little sensibility as to what was required by their underlings because it looked easy to say, "we just need to create the place, and they will come" only to find after some time, it was not the case. Although small groups did work with the tools at hand, a large portion of the company did not contribute 100% to your system. The general question evolved from centralizing, to distributing, to searching, to portals, on and on. It looked to me like a never ending tail. All this tools where being created first, and then we where supposed to find the ideal use out of them.

A nice example of what happens when you hit on people's needs is the
Apple's Ipod , although not the first musical device in the market , it was the first to tap into people's need to listen and enjoy their entire set of music in a small form factor with a simple to use interface and elegant design. The key to its design was or is how

---

[5] Other acronyms with more or less complexity are DRM, ECM, BPI, etc. They all do the same basic functions.

intuitive (no "thinking" needed) the controls where; I purchased one of the first versions the next day a friend showed me his. I was frustrated with all the other "solutions" because they were cumbersome, could not store more than a couple of songs, and I was imagining a solution to include some kind of hard-disk. The actual music problem was the headphones quality, and I knew right there the ipod would be very successful.

One thing that Apple decided to do was to leave certain parts of the device "open".
Ipod users then came up with new ideas on how to take advantage and create new media formats like Pod Casts (Now a common word) and a ever expanding range of accessories. We can see it everywhere. That is the effect of tapping into people's needs and a good example on why customers needs (if you want to innovate) are the key to succeed.

"Ideas generated by lead users at 3M were not only more novel, with a much greater potential for revenue creation, but they also were found to "address more original or newer customer needs, to have significantly higher market share, to have greater potential to develop into an entire product line, and to be more strategically important," writes von Hippel in *Democratizing Innovation (MIT Press).*
When users have no "constraints" to come up with new solutions to a problem and apply them to existing products, we can witness the powerful benefits when those ideas are embedded in the subsequent version. Users are the direct beneficiaries of innovation.
Imagine giving the same space and authority to your own employees. (See IDEO section). Google shows what this strategy can do for business reputation and profitability by allowing 25% of their (employee's) time to develop pet projects.

It is therefore my experience and conclusion that attempting knowledge "control" will impede the creativity and collaboration you need to put any organization ahead in the game of innovation.

The third and most important factor to KM troubles, is not understanding what knowledge is. Knowledge concepts today are not something we can understand easily in two phrases ; it is why this book was written; to try finding a clear concept of what knowledge is so that we can be clear as to what is possible to design and create.

Summary:

We can conclude that people are always at the center of any knowledge project, hence the human factor is always the key ingredient to its success.

In the following pages you will find new ways of looking at knowledge and points of view that I hope can help you discover how we can re-direct our concepts into a better framework, new problem solving mentality and reorganize priorities to drive success. You will find reference to a new leadership focus, and a solution designed to allow us to work collaboratively; the opportunities to do it are increasing fast in this new Internet era

Chapter 1

What is the problem?

 When we look at a problem in today's society, many times we are blinded to the actual problem by the symptoms. Take obesity for example, theories go from lack of exercise, to amounts of food, etc. If that was true, the problem would have been solved long ago.

The first sign that told me there was a more serious problem underneath, was when I started to listen (I always had a hard time doing so) to people when asking for my help. I kept hearing over and over the problem explained with technical language, and tools.

Example:

Customer: "Yes, We would like to know about your experience in X software".

Me: " Ok, I've been working with it for ten years, but first I would like to know what is the problem you are trying to solve?"

Customer: " We are looking for people who knows about X software to implement it here"

Me: "Right, but what is the problem that you are trying to fix with the technology?"

Customer: "Oh, we need to create a website or document repository where people can store their documents"

Me: "Does that mean that currently you can not easily find those documents, correct?"

Customer: "Correct, we need to be able to access the information that is not easily available"

Me: "How do you plan to convince people to post their knowledge in those documents on a regular basis?"

Customer: " We hope they can see the benefit once is implemented, that is why we need X software"

Me: "Do you have policies in place or a strategy for motivating people into sharing their knowledge?"

Customer: "That is what we need you to come and help us with"

Although they could "see" they had an issue with documents, the real problem was behind that.

The customer would get a little frustrated, because she could not "see" the problem with her logical thinking. And I did not have the skills to explain it simple and clear either.

She or He was connecting **problems** to **tools**, expecting the tool to equal the solution. Before defining the whole problem, some software tool was included in the conversation.

It made me realize how IT projects are in trouble from the get go because we believe technology = solution.

Philosophy and Religion

The area that studies the concept of knowledge in academia is called *epistemology,* a branch within philosophy; the definition I found about epistemology is -- tries to explain what it is, its nature, origin, and its scope (of knowledge).
Coming from an engineering background, I found very complicated to understand their philosophical material (See definitions section). . Why? Because we engineers tend to look for simple explanations to a concept, so that we can build a product with the tools at hand. I agree that we make things complex, but in order to build a plane, you need to understand the concept of flying.

When thinking about this book, I found that the only way to make it readable was to come up with a solid reasoning base, find or create simple explanations about knowledge and focus it into practical software solutions and avoid complicated arguments as to what knowledge is.
Because the concept of knowledge is and has been a very contentious topic in religious and philosophical circles for ages, we'll look more into it in the next sections. But first, lets look at the why question.

Why do KM Projects fail?

The fundamental causes of failure are:

Leadership

Leadership has always played a key role in the success or failure of any project in history. In IT and especially in knowledge management projects, there is primarily been a lack of understanding about how to properly envision, plan and execute a knowledge solution. It includes not understanding what knowledge is or means and where does it reside. The way it has been done is the same as any other IT project.. In reality each IT project is very unique, and statistics show the high percentage failure rate IT projects have due to misconceptions from the get go. The fact that we had always gone trough difficult times doing these type of projects proves how complicated we make them without understanding the basic and fundamental drivers/concepts/causes for the business. It applies to any field. If you want to fly, you need to understand the concept first.

Added to the mix, is the aggressive vendor sale pitch from software companies trying to sell a product to you. "The mouse trap" as someone said to me.
There are many reasons, but the biggest sin leaders/managers commit is thinking technology IS a solution (Go to Views and Concepts) as seen in the earlier conversation. I witnessed it when there was a project trying to re-create an in-house developed software package. When the new team members looked into the details, it was very hard for them to come up with the concept as to why certain functionality was done in a particular way. In many cases after asking colleagues, it is better to scrap the product entirely and start from scratch. Working with software already build most of the time is a waste of time for developers. They prefer to start with a fresh view, and try to enhance the prior software with new and more intuitive functionality.

On the other side there is a poor conception of the human factor as mentioned earlier; understanding that users day to day job will be affected by knowledge based systems helps build a better answer. There is a big misunderstanding of what the impact is when you communicate with people through training exercises, planning, and execution about a new software system. I refer to how early do people (end users of the system which in KM parlance is literally everyone) get involved with the project to get their BUY IN, how much decision power they have to define the details (Normally none) and how their valuable input is included in the project and training plan. Why? Simply because they are the ultimate beneficiaries of your project. The tasks that involve users are the most important. Training is the most underestimated effort I have seen in most knowledge projects. **Training material and people are the key to a successful outcome** for a knowledge project. In many cases is about bringing and installing a software package, configuring it, and then conducting a simple training for people on all the features the software has, with no correlation to their day to day work and how they should apply it; What are all the benefits for them and the company by using X,Y or Z package?

There are other cases where the opposite is true. Leaders realize they do not know, and create an exercise to ask the users what they want or for details about how the project should focus based on their needs. Example: asking users how the training material should be created. Although in essence it sounds good practice, the resulting reality is getting different opinions from users that do not align to a global goal makes it harder, because each user has a different opinion as to how things should be done. So both extremes are bad, and the right balance is what will help you succeed in the long term.

Not understanding the big picture of why do you need KM or knowledge strategy and not asking enough questions about "What is the fundamental problem?", where do we want to go?, why are we doing it?, What are the rules and processes we need in order to accomplish our goal? What do we envision? What is OUR solution? What is our fundamental goal? etc is what we want to help you answer.

The above is an intrinsical task for the leadership and top management group in your company. I've learned about some theories and practice that look at opposite skills from today's understanding of what leadership should be.

"Show me a hero and I'll tell you a tragedy." F. Scott Fitzgerald.
A book by Joseph L. Badaracco "Leading Quietly: An Unorthodox Guide to Doing the Right Thing" looks at the opposite traits of leadership: restraint, modesty and tenacity shows us that what we see today might not be the ideal way to leadership. -If you look behind lots of great heroic leaders, you find them doing lots of quiet, patient work themselves. —
Leadership plays a key role in driving success through learning. (See Learning Organization)

Unreasonable goals

Expecting end users will understand everything on their own, without a clear connection to the business purpose is a sure way to hit the wall.

The transition from paper based processes to digital has posed another big challenge for leaders, and old management traditions many times prevail. When we tried to implement a workflow (Digital process) in a project, we found a lot of resistance from not only users or low level employees but managers as well. Why? Because the reason (In their eyes) they were hired was precisely to do process monitoring (Some may not be aware of it), follow up and control getting things done; suddenly it was the opposite, their work could be controlled by an application, so when the process is no longer under their control so to speak, they feel (understandably) threatened, because now you only need a click of a mouse to make the process happen, somehow making managers feel they are displaced without their signature. The more bureaucratic the culture, the bigger this syndrome is. The other part about digital processes is that they can reveal the truth in terms of the real demand, status, delays and managers might feel overwhelmed with the amount of work generated by becoming "Paperless" along with the pressure to complete assigned processes; we made the process easier to generate, but managers find themselves in hot water trying to juggle the additional work load. A project well understood, planned and communicated on digital processes should actually help lower inefficient work, allowing you and your managers to concentrate and spend your valuable time in more important and strategic areas of your business.

Carelessness can create such goals like "We want to go paperless in 5 months". Paperless in reality means that not only your company, but every other company you have contact with has to change their methods so that they communicate with yours. Unless you have a Walmart type of business, it is difficult to accomplish.
"Managers complain that people frequently hoard knowledge, fail to share it with others, and generally behave uncooperatively. In high-stress organizations", says Brian Bacon, "ninety percent of the people don't say what they mean or do what they say. Which is why ninety percent of what should happen, doesn't happen."

Psychology

"When human beings are placed in narrowly defined positions where those above control the fate of those below, the effect is to constrain our intelligence, responsibility, and creativity" Elizabeth Debold, The Business of Saving the World, 2006[6]

Ten or twenty years ago (and even today), we were/are told information (and knowledge) is power. The problem with power, as with money, is that everyone wants to accumulate more and more, having a hard time detaching from it when we tell them the opposite " Please share your knowledge here". *Marc Dotson* **published May 30, 2006:** "United 93's lesson yields knowledge is power "Knowledge is power, and pure knowledge must unite us.". The more we send this type of message to people, the harder the goal will be to share knowledge. Equaling knowledge to power, evidently will make people hold on to it, instead of sharing it, which is the key to collaboration and team work, which in turn is the key to creativity and innovation[7]. A twist to the above quote could be

---

[6] http://www.wie.org/j28/business.asp?page=6
[7] Innovation: Is the introduction or new ideas, goods, services and practices which are intended to be useful.

"Knowledge power depends proportionally on our capacity to unite ourselves collaboratively" and would allow us to bring more easily people to share their knowledge producing greater results.

The result can not be more obvious: People will resist and even subvert to give away what they know without a clear tradeoff and understanding for doing so. In this book I am joining the theory about how to motivate people into sharing their knowledge (

Unclear Concepts About Knowledge and Tools

As I stepped back to think about KM, I asked my self what was the meaning of knowledge. I found confusion in some people's definitions when I asked friends and colleagues about what the word knowledge meant to them. Amazed by the discovery (disbelief) with the lack of understanding by me of their dictionary meaning, I decided to be more disciplined and always check for their meaning before attempting a meaningful debate.

If you decide not to pay attention to this fact (i.e, dictionary as point of reference for the meaning of words), even trying to agree on what the problem is or it's possible solutions will be almost impossible; with no common denominator for basic words definition it is unlikely you'll succeed and is exactly what happens in many IT projects around the world; because the words used to define the basics (Causes, symptoms, consequences, solutions) will have a different meaning for each team member in the group, consequently confusion and chaos will follow. The key to a successful start is based on common definitions understood by all involved; how well the team understands them will enable clear understanding of the problem, the solution, and then the tools to use. (e.g., **Bleak outlook for IT outsourcing** : Fundamental misunderstandings between government outsourcers and suppliers mean most projects are 'doomed before they begin', a survey has found [8])

---

[8] http://news.zdnet.co.uk/business/management/0,39020654,39273622,00.htm

Any challenge we encounter in life is a problem to solve. We need a successful problem solving thinking method to consistently do it right. Creating and running companies is no different.

"Instead of thinking of ways to create a business, entrepreneurs should be thinking of ways to create value through solving a problem. And that is a deep, thorough process. "Half of all the energy we spend in inventing new business solutions is first figuring out the problem." Jay Walker inventor of Priceline.com and holder of more than 100 patents.

Putting a company underneath the solution to a problem is the easier part. As Susan Nichols found when her sweaty feet slipped during a Yoga pose, made her go looking for a fast drying alternative after a rug she bought did not fit her need causing a nasty fall. After finding nothing in the market, she decided to make her own. She stumbled upon a water bowl for dogs with rubber nubs on the bottom to prevent sliding. "That's it", she thought, figuring out the nubs could be affixed to micro fiber for added grip. She discovered a Korean factory that could manufacture it, then in 2003 she founded Yogitoes to sell her skidless towels. The pay off, in two years she went from selling one million to three million.

If not only the leaders, but the rest of the team believes technology is a "solution", then they are bound to fail in the long term. (Described in next section)

Discovering people believe technology = solution was shocking to say the least. I thought the same way until I've got that Aha! moment. I discovered the pattern when sitting in various IT meetings, and paying close attention to people's words about what the technology could do to solve the problem. That was the first time I realized it was a backward thinking, instead of agreeing what the problem was in the first place and coming up with the solution in **non technological terms**. By stepping back, it sounded confusing technical language being used in defining the problem. Later, I decided to ask about what was the fundamental goal for the various knowledge projects to Christine, after a short pause she would say to me like realizing something deeper: "I do not know". It was not clear why they were trying to implement all those different applications. I calmly summarized it: "The problem is you can not find the information". That was the symptom. The core problem was convincing people to post their knowledge. So bottom line: It is always about people. If people do not want to share their information, no technology miracle will force them. (See appendix 4 for the possible solution)

In another meeting, someone launched a question (maybe in desperation) that it would help me start realizing many things in this book. I was finding out new insights by listening and observing people, then stepping back to reflect about it.

Industry fuzziness: A case to show evidence of how we build complexity (unnecessarily) into everything we do.

There is an inherent problem with us technology people and specially the ones that have to sell it. They need to have a new buzz word (i.e., Sales pitch) every year or so to make sure they can "wow" their customers into a new sale (albeit being the same with more stuff added). This is good for businesses bottom line but terrible for effective practice, and end users buy in. The products get tweaked or enhanced not always in response to customers, but to make the sale ; applications become more complex with each iteration. In many cases, companies are known to have 20-30 different software applications to deal with information and processes, which does not make sense to regular business people.

The other reason KM projects keep failing is because the fuzziness and lack of clarity software "solutions" provide today. Software (excluding open source) does not include the fundamental concepts as to why it was built in a certain way specially the language chosen to name features. Without it, when the time comes to make changes to adjust to corporate culture, is even more difficult to understand what the implications are. Software companies' answer is to modularize changes as the way companies can enhance its solution, without realizing the long term impact.

Here is an example and list of topics of a content management convention I found happening in 2006:

"• Content Management Architecture, Infrastructure, and Applications
• Data, Image, and Content Capture: The On-Ramp to an ECM Solution
• Centralizing Your Assets: Store, Retrieve, and Preserve
• Communicate, Collaborate, and Manage - "Can You Hear Me Now?"
• Making the Invisible Visible: Measure, Monitor, and Analyze

The fourth topic that talks about Collaboration is interesting. It is astounding the number of elements I found on each topic in their webpage:

• This theme provides advice, insight, and discussions about how to best share, route, deliver, secure, and control enterprise content so attendees can meet or exceed their compliance, operational, and performance objectives. Areas of particular focus include: email, Instant messaging, Web content, and Portals. Other sources of enterprise information and best practices for process and policy analysis are addressed.

.

- BPM
- Workflow
- Document management
- Email management
- Instant messaging
- DRM
- WCM
- Collaboration tools
- DAM
- Security
- Localization
- Syndication
- Personalization
- Publish
- Portals
- Case studies
- Versioning capabilities & controls
- Process planning
- Case/exception handling systems
- Cultural/organizational/ownership challenges
- Unlocking valuable information trapped in transactional systems
- Building a performance-accountable organization"

It might be seen as useful software, but you tell me if it is confusing or not, it is for me even knowing what each of those are supposed to do. Many different pieces and no simple way to understand how all of them work in conjunction, what is the whole picture, what do they accomplish, what is the problem, how should we start and no basic concepts behind them as to why they were built like that, and in general a one-solution-fits-all approach. We are stumbling again and again trying to connect a problem straight to a tool believing that's the answer. If you want to build a house, the hammer and nails are not your solution, ; the **design** on a paper is. The tools allow you to build it. In business terms, it is the interface design, processes and policies document, views and reports, etc that define your solution **to be built**.

How this misunderstanding affect IT Consultants?

Technology consultants

• We consultants should be aware of not mixing technology with solutions. But I found that many fail to raise the red flag by falling into the trap and giving wrong answers to an undefined problem and failing to ask many questions from the very beginning; Why did you decided for X software? What was the criterion? What is the leadership vision? What is the correlation from that vision to this project? The main reason: I would say is mostly fear. Fear of being direct and honest, fear of loosing a customer given how competitive the market is, trying to look good in front of a customer, boss, etc; when consultants are not free to say what they think and make their own calls even dropping projects for their lack of definition, they try to give answers to something that is not clear and that has not been defined as the core problem; more troubles will arise, either in the short term or in the long run..

Many projects I've seen and experienced are ill conceived from the very beginning. Once a consultant is brought in, the project is already bound to fail due to lack of knowledge or understanding from leadership. By not involving consultants from the very first steps when planning a project and its outcome, is a confirmation as to why the failure rate in IT projects can be as high as 70% to 80%. Deborah Weiss, Meta Group program director of enterprise planning and architecture strategies, said up to 72 percent of all IT projects are late, over budget, lack functionality or are never delivered as planned. Many authors have different opinions as to why projects fail; the common element I found on all reports was the lack of a clear problem definition plus a solution design which defines your metrics for success. Projects that are ill defined will most likely fail. Understanding what the core problem is (using PSST) might be the light IT managers have been missing for a long time.

Ethics are also affected when there is strong financial pressure to acquire clients and keep them "happy"; consultants dread having to say no or lack the ability to guide customers into a better **systems thinking (See Appendix 6)**, methodology; saying clearly "I do not understand the why", we continue working in projects that are highly risky, prone to failure, and many times working in situations that are deadlocked. For example:

Consultant to customer: "What do you need?"

Customer: "I do not know, show me what you've got" or "when I see it, I'll know".
Effectiveness: Making sure you check all fuzzy words used in defining something by the customer provides you a base for better communication and less errors. I can't stress how important this is. The meaning of the words used to define a project is as important as the project itself.

The consultant's role has been deteriorating over time, from experts who could tell you right from wrong at any time, to yes people who would follow commands regardless of their logic. It happens to me all the time. "Here is the software, please help us implement it". See notes "The outsourcing effect". A good practice about how information flows can be found in the next topic.

Connecting the dots. A-B-C-D-E

A. "In the long run, the only sustainable source of competitive advantage is your organization's ability to learn faster than its competition."
B. Your Ability to learn lies in the ability to nurture Team Learning
C. "Team Learning is vital because teams, not individuals, are the fundamental learning unit in modern organizations." (See Chapter 3)
D. Collaboration is the key for Team learning through technology.

New technology is needed to actively promote knowledge in organizational teams.

Figure 1. Connecting the dots

What is the Solution? (See Chapter 2)

- Having an Open Mind.

- Use Clear Concepts

- Implement New Leadership abilities

    - Different set of priorities
    - L-V-S-P-R-S
- Learning Organization (See Chapter 3)
- A new technological platform (See Chapter 4)
  Your major goal: The main goal is to achieve 100%

  collaboration

Chapter 2

The solution

1. Having an Open Mind

It is very important to understand that before education, having an open mind every day for the rest of our lives is the key, teach your children about it, because there are cases when highly educated individuals tend to close their mind to new information later in life because they think they already "know enough". Keeping an open mind through out our entire life is the first step into true learning. It prevents culture, beliefs, religion to interfere with our learning. The mind is like a parachute, it works only when it is open. Wise people tell you "The more I know and learn, the more I realize what I do not know". I come from a country with many problems in the surface, but as a report found recently (), Colombia ranks in the group of happiest people in the world. The reason is, we are open minded, and people with an open heart; we believe in peace within even though outside the world around seems to be ending.

2. Use clear concepts

I realized the need to change our mentality in order to find the REAL reasons of why a problem occur as doctors do within the medical field; they look at the external *symptoms*, but never conclude that the symptom is in fact the problem; they always look for other symptoms to find what the hidden cause is (illness) generating the symptoms. (To my amazement, this is what I found to be the original idea behind the scientific method) See Edwin Smith Papyrus (circa 1600BC). If I told you that the obesity epidemic in the USA has nothing to do with diets, but with TV, what would you say? (See article[9]). Most people reject the idea because they cannot "see" the hidden link or problem. This is exactly the issue of not thinking systemically.

Most situations in business, what you see directly are symptoms rather than the cause of a problem, a sign that something else is going on. As in the medical field, if doctors do not find the root issue —illness-- that produce the symptoms that make you unhealthy, it is unlikely they will find a cure for it. The problem will remain as long as doctors treat the symptoms as the main

---

[9] http://www.sciam.com/print_version.cfm?articleID=0005339B-A694-1CC5-B4A8809EC588EEDF

problem[10]. In many cases, not only the problem does not get solved with the intended solution, but it gets worse either generating new problems or becoming a bigger one on its own.

One consequence of this short term (un-systemic) thinking in the technology world is the number of applications we have to deal with today that do not solve the core problem with information and knowledge in general. For each symptom, there is a different application that tries to solve it directly. The result: a huge list of software applications for every imaginable symptom.

In the worst case scenario, you can end up solving problems that people don't want solved. For example, In Scandinavia, the wire cheese slice is a household staple. To OXO International (A design company), engineers it had an obvious flaw. Consumers use the slicers daily, the wire eventually breaks and needs to be replaced. The engineers remedy was to incorporate a blade instead. But when OXO tested the product in Denmark, customers weren't interested. "The wire slicer is the tradition," says category director Michelle Sohn. "Even though it has a problem built into it, they kind of like the problem." So instead, OXO is working on a more durable wire.

Same happens in the IT world. When good solutions are created, they are not durable (Live long), and are susceptible to be replaced suddenly, as it happened to a former customer where the DMS[11] software vendor declared bankruptcy, forcing them to search for another product, generating more problems while doing so –training, new environment, users complains, etc—because the dependence customer created on software vendors. The story tells us, many years ago, the customer had a software package that handled in an effective way all documents in a central repository. The software company went bankrupt, forcing the customer to search for another software product, without realizing all the problems they were about to face: Different technologies, different concepts, different look and feel that would affect all users (The new product was web based, the old was not) and simply put, a very different way to work, that they did not anticipate. The negative long term effects were various. Users complained about not been able to find documents and information, with multiple systems storing the data, but the biggest effect was user perception and more importantly their "trust" in technology was affected because the new system was unreliable, did not perform as expected, etc, etc. Users rejected it and did not want to deal with anything different than what was in their desktop PC even

---

[10] This explains P. Senge famous phrase "Todays problems are yesterdays solution".

[11] DMS Stands for Document Management System, a database where users store centrally all their documents.

thought the PC had a higher crash rate than the other system, and users would loose everything.

Finding the Core problem:

Before Google appeared in 1998, there were eight or nine different search engines competing for the top spot. The search problem in their thinking was not about which pages where the most relevant, but who paid the most money to be listed in their results. Sergey and Larry finally found the ideal solution to search: The value of ranking information was in the relations (hyperlinks) between pages, so counting them would give a better measure as to what people considered an important resource or page. They realized that to make it work, advertising must be excluded from or be independent of these results. Today, Google sells six *billion* dollars in ad words by coming up with a powerful yet simple solution.

There is a *practice* of implementing software tools to create, organize, control and share information (DMS, ECM, BI, search, data mining, etc), but leadership in various corporations does not realize that people are an intrinsical part (And the most important) of knowledge based systems; and people need to be genuinely motivated to post, share and collaborate with your software "solution" in order for you to achieve your goals.
We can see now powerful collective platforms like Wikipedia and Digg that are creating collaborative knowledge that's far better than any editorial staff could make it to be. (See Wired News [12])

From observing Wikipedia definition page for knowledge, I can say that culture and education plays a big role into having an open or closed mind to new ideas.

Today's Knowledge definitions

As I stepped back in order to understand, I came down to ask my self: "What on earth is knowledge in the first place". I looked in various dictionaries and encyclopedias (here I list what I found). I learned that philosophy studies the term through epistemology. So I tried to search for a practical definition that I could feel more connected to.

Here is an example knowledge in philosophy:.

"An attempt to develop a theory of knowledge and a philosophy of mind using ideas derived from the mathematical theory of

---

[12] http://www.wired.com/wired/archive/14.07/people.html

communication developed by Claude Shannon. Information is seen as an objective commodity defined by the dependency relations between distinct events. Knowledge is then analyzed as information caused belief. Perception is the delivery of information caused belief. Perception is the delivery of information in analog form (experience) for conceptual utilization by cognitive mechanisms. The final chapters attempt to develop a theory of meaning (or belief content) by viewing meaning as a certain kind of information-carrying role." *Fred I. Dretske*

To my mind, it is not simple, neither clear as to what is trying to say.

For me, it was striking to find this out; as no one I talked to was very sure of themselves answering what knowledge meant. We all seem to assume we "understand", but looking closely at dictionaries and encyclopedias, I found some inconsistencies or gaps that needed attention; also they did not connect to my practice in dealing with KM. The more I asked the question around, the more I realized there was no clear understanding of the word knowledge , and that I had to find a common ground for reasoning to come up with a view understandable to anybody.

Definition From Britannica:

"a posteriori knowledge:
knowledge derived from experience, as opposed to a priori knowledge (q.v.).
a priori knowledge:
in Western philosophy since the time of Immanuel Kant, knowledge that is independent of all particular experiences, as opposed to a posteriori knowledge, which derives from experience alone. Royal Society
the oldest scientific society in Great Britain and one of the oldest in Europe, founded in 1660. It began earlier with small, informal groups, who met periodically to discuss scientific subjects.
Knowledge
from the salvation *article*
Religions that trace the ills of man's present condition to some form of primordial error, or ignorance, offer knowledge that will ensure salvation. Such knowledge is of an esoteric kind and is ...
Transmitting knowledge
from the technology, history of *article*
In the ancient world, technological knowledge was transmitted by traders, who went out in search of tin and other commodities, and by craftsmen in metal, stone,

leather, and the other mediums, who ...
Self-knowledge
from the Locke, John *article*
Some ideas are not of things outside the mind but are reflexive and internal. Locke finds it necessary to classify these in Book II and in doing so sets down the foundations of empirical psychology"

From Dictionary.com

knowledge n : the psychological result of perception and learning and reasoning [syn: cognition, noesis])
Pronunciation Key (nlj)
n. The state or fact of knowing. Familiarity, awareness, or understanding gained through **experience or study**. The sum or range of what has been perceived, discovered, or learned. Learning; erudition: teachers of great knowledge. Specific information about something. Carnal knowledge.

This definition was the one that caught my eye first. It found the two main elements I experienced at work. Study, or information, data and experience (processes, tasks, etc). What I did not agree with was the OR in the middle. Then I continued searching.

Knowledge is the confident understanding of a subject, potentially with the ability to use it for a specific purpose. The ability to know something is a central (and controversial) part of philosophy and has its own branch, *epistemology*. On a more practical level, knowledge is commonly shared by groups of people and in this context it can be manipulated and managed in various ways."

I liked this definition better because there was a better correlation between theory and practice (What I was doing). I combined the two previous definitions to build a new simpler concept. A concept involving the elements data, processes and intellect.

While working and reading on other subjects, I stumbled upon something that I was sure would help.
Timothy Williamson, in his book Knowledge and its Limits, seeks to revert the traditional conceptual priority of belief over knowledge, instead seeing belief as dependent on knowledge. It helped clarify one of the biggest points and arguments in philosophy, when trying to explain knowledge. I could not agree more with his view.

In my experience, belief[13] is dependent on knowledge and the state of mind you choose, positive or negative. By state of mind, I refer to the percentage of negative or positive thoughts you let float in it. If you are or tend to be a negative person, you choose to produce many negative thoughts which are closed in nature; you'll start slowly believing and accepting only what you know and believe, oblivious to anything external that tries to point in the opposite direction , therefore your reality gets distorted into a negative spiral. Proof of this can be found in the book and practice seminars called CMR or cellular memory reprogramming. It explains that the neurons in our brain function like small magnets, attracting those with the same nature. Negatives with negatives, positives with positives, generating a chain reaction explaining the common phrase "The more negative you are, the more negative thoughts you tend to attract, the more pessimistic is your attitude and outlook. Same in the opposite side; Eastern practices like Yoga and meditation, helps you find a balance in your mind through specific exercises, so that you do not tilt one way or another. The ideal state is to be in a neutral (no thoughts) position.

If in the opposite side you are a positive person consistently, open in nature, then your thoughts, attitude, and actions will always look for positive outcomes which you'll believe true as well. "All of our feelings, beliefs and knowledge are based on our internal thoughts, both conscious and subconscious." Donald Martin 1991.

How can we apply these concepts into technology? What are we missing that makes it so hard to bring people into sharing their knowledge?

The problem we have today in IT is that these definitions do not translate easily into bits and bytes (Programs) and into a practical and simple concept we can build tools from. I needed a simpler way to understand it to make it practical.
Still, the philosophical hurdle made me to keep looking for a way to talk to people in philosophy circles. I learned the hard way. After meeting a very nice lady with a physics doctorate background combined with a deep philosophical mind, we immersed in an intense debate about knowledge, mind, belief, etc, etc, which I can say did not go well the first time. I was saved by my baby daughter asking to go to play with her. But the experience made me even more resolute to search for the answer. While reading a project management book I discovered about the Occam Razor principle. I certainly could not believe it. To me, philosophy material was complex to digest and understand, yet to find a philosopher suggesting to do the opposite was unexpected.

---

[13] See Appendix 2

Parsimony Principle

The principle *"Pluralitas non est ponenda sine necessitate."* --- written by William of Occam— from the twelve century philosopher. The principle of parsimony, also known as Occam's Razor, translates from the Latin as pluralities should not be multiplied without necessity. It means that truth is most often simple and lies are most often complex. Parsimony states that when two or more theories are presented that propose to solve the same problem, the simplest that gives the most complete and satisfying explanation has priority to be studied in detail as it is the most probable. If during study the simplest is found to be flawed, then it is excluded, multiplying pluralities becomes necessary, the next simplest gains priority, and so on.

"He believed that people often add complexity to situations even though it doesn't help to resolve them. He suggested that the best way to figure out things out was to find the simplest explanation and use that first because, most of the time, it was the right explanation. "Scott Berkun

"If you find yourself in a complicated problem, open yourself to the possibility of a simple solution." SignsOfLight.org

The word knowledge is one of the most contentious topics people debate on today (and for many thousand years past), and it seems like a never ending discussion.

I found that before explaining anything about this book, I had to drive people to a common point of agreement to start from. Without it, discussions would take hours and never reach an end. To be in agreement from a common starting point ,we need to differentiate good ideas from bad ideas, what is right from what is wrong,.

A Base for Reasoning: The principle of good

"All our ideas should produce good and lasting results and then anything that is good now would have been good in the past and it will be good in the future and it will be good under any circumstances, so any idea that does not cover all this broad base IS NO GOOD.

The principle of right

To be right, one's thought will have to be based on natural facts, for really, Mother Nature ONLY can tell what is right and what is wrong and the way that things should be.

My definition of right is that right is anything in nature that exists without artificial modification and all the others are wrong.

Now suppose you would say it is wrong. In that case, I would say YOU are wrong yourself because you came into this world through natural circumstances that YOU HAD NOTHING TO DO WITH and so as long as such a thing exists as yourself, I am right and you are wrong.
Only those are right whose thoughts are BASED on natural facts and inclinations."(Edward Leedskalnin, "A book in every home", 1936)

Where to start

A different way to start exploring what is knowledge, is to ask yourself where does it reside following nature's architecture of ourselves; manager surveys as of 2006 report "80 % of corporate knowledge is in peoples minds".
The brain is like the CPU in computers, and the mind is like software running on top. The brain is divided into two hemispheres (Fink, Marshall, Nature vol 382, p 626). Right and left.
Consequently it should indicate that there are two types of knowledge for each side and how similar they are. The right side of the brain handles *intuitive knowledge* (Mona Lisa Schulz Md Phd, Awakening intuition - April 20, 1998), while the left side handles *rational knowledge*. (Fink, Marshall). By analyzing where knowledge resides we can easily understand better how knowledge is divided, then we can look at the definitions and elements on each type... Mona Lisa Schulz Md Phd, explains also in her book, how the body is an Intuitive knowledge channel (Paperback)[14]
In applied Kinesiology, it talks about a method that allows you to access knowledge through our own body.

---

[14] http://www.amazon.com/Awakening-Intuition-Mind-Body-Network-Insight/dp/0609804243

Right Brain vs. Left Brain

The structure and functions of the mind (Fink, Marshall) show two different sides of the brain control two different "modes" of thinking. It also suggests that each of us prefers one mode over the other. Men tend to be use the left side , while women tend to the right side. Modern society (Western) educates children focusing in developing only the left side (Intellect) over the right side (Humanistic, arts).
Experimentation has shown that the two different sides, or hemispheres, of the brain are responsible for different manners of thinking. The following table illustrates the differences between left-brain and right-brain thinking:

Left Brain	Right Brain
Logical	Random
Sequential	Intuitive
Rational	Holistic
Analytical	Synthesizing
Objective	Subjective
Looks at parts	Looks at wholes

Some people , however, are more whole-brained and equally adept at both modes and many cases researches have found geniuses to be the ones using both sides and in savant autistic (See Daniel story below) cases. In general, schools tend to favor left-brain modes of thinking, while downplaying the right-brain ones. Left-brain scholastic subjects focus on logical thinking, analysis, and accuracy. Right-brained subjects, on the other hand, focus on aesthetics, feeling, design and creativity."
(www.funderstanding.com, 2005)
Intuitive knowledge (see section below): Intuitive knowledge is viewed as insight that comes directly in dreams/images and "gut" feelings; those dreams are made of images. Here are some excerpts from the Internet; Intuition: way of knowing directly; immediate apprehension. The Greeks understood intuition to be the grasp of universal principles by the intelligence (nous), as distinguished from the fleeting impressions of the senses. The distinction used by the Greeks implied the superiority of intellectual intuitions over information received by the senses. Christian thinkers made a distinction between intuitive and discursive knowledge: God and angels know directly (intuitively) what men reach by reasoning. A person who has an intuitive opinion can not fully explain why he or she holds that view. Intuition is an unconscious form of knowledge. It is immediate and not open to rational/analytical thought processes. It differs from instinct, which does not have the experience element. An important intuitive

method is brainstorming.

Rational Knowledge

"Knowledge is information combined with experience, context, interpretation, and reflection. It is a high-value form of information that is ready to apply to decisions and actions." T. Davenport et al., 1998 "the insights, understandings, and practical know-how that

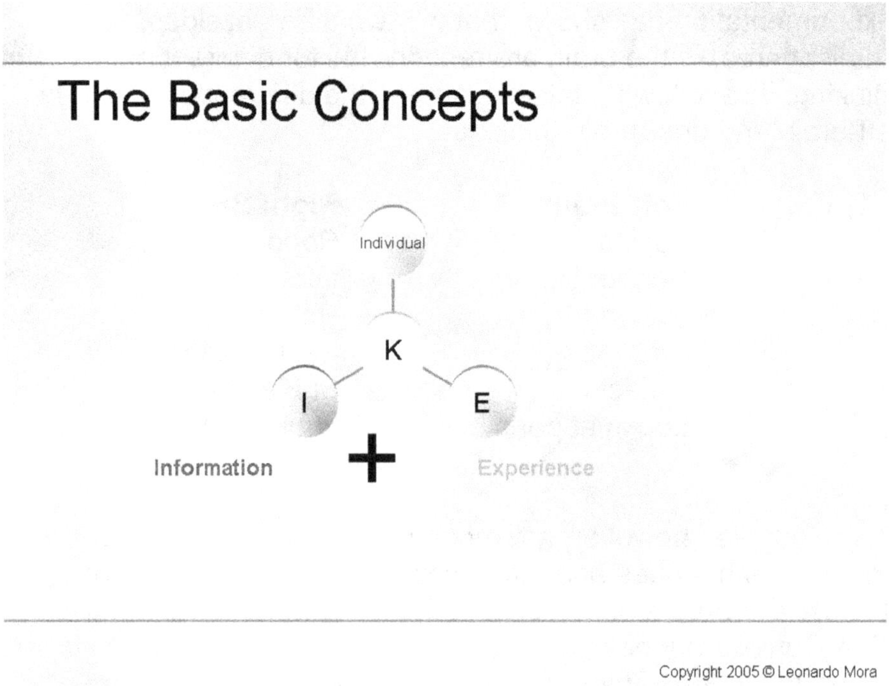

The Basic Concepts

Individual

K

I   E

+

Information   Experience

Figure 2 The Concept

we all possess – is a fundamental resource that allows us to function intelligently." Wiig, 1996.

Based on my work in software technology I came to the conclusion that in very simple and practical terms, rational knowledge is indeed INFORMATION plus (+) EXPERIENCE at the individual level. The way we learn since we are born comes in the opposite order. Experience first, then we learn how to read which takes us to learning how to study to gain information. Note that there is a plus in between the two words, this is the opposite to the current notion that it is insight gained through information (Study) OR experience; If you read all the books (Information) about riding a bicycle, it will tell you all about how to do it, but until you try, you only have information about it. Having the ability to perform is what constitutes knowledge. Having information only does not mean you completely understand how to do something (experience). Once you do it, then you understand the meaning and the difficulty of the process. This applies to any other field and is the basis of what is known to many as the Scientific Method (See below).

Experience (or practice): means doing, going through a defined process (which by definition is a series of steps or tasks) from start to finish. One key aspect of positive experience is that you have to be willing to do it completely in order to gain full insight (A partially completed process does not mean you fully experienced something). Some times a bad experience like a robbery or an accident is a sudden event and not a planned process. I define them as negative experiences that we are "forced" to live in our life. Certainly, if we are made aware that an accident or any negative experience is likely to occur in the near future, we are not willing in most cases to go through it. You might ask about high risk sports like formula racing. In that case, pilots have trained extensively, and have the skills to minimize the chances sudden accidents, although not completely. We accept the risk based on our skills. Anyone can tell you that even attempting to drive a formula 1 car would be very risky, so you should not even try unless you get proper training and experience.

Experience is the portion of knowledge that cannot be easily transferred or taught, and is not easily detachable from the person as Leadbeater (2000: 70) argues "This sort of know-how cannot be simply transmitted. It has to be engaged with, talking about and embedded in **organizational structures and strategies. It has to become people's own.**" Or as Professor Ikujiro Nonaka explains,

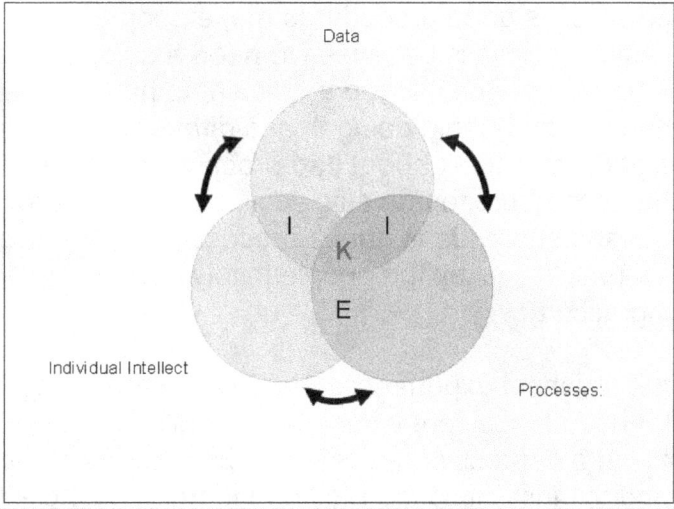

Figure 3 Main elements

"Tacit knowledge(*Experience*) is personal, context-specific, and therefore hard to formalize and communicate. Explicit knowledge (*Information*), on the other hand, is transmittable in formal and systematic language. Without experience, we cannot truly understand" He also refers to how personal

knowledge becomes organizational knowledge, but I add to it (only through team work and collaboration, as we will see later):.

Learning (See Generating Knowledge) through experience is one the strongest ways of gaining new insights as it involves all our senses.

"In plain words, rational perception entails accepting and believing only what is true based on one's study, observation and experience. The knowledge derived through such an approach is rational knowledge. It should be pointed out that the process of attaining rational knowledge begins by having an open mind and by giving up blind faith and notions prevalent in society", by Jainism.

Here are some definitions from www.dictionary.com for experience:
The apprehension of an object, thought, or emotion through the senses or mind: a child's first experience of snow. Active participation in events or activities, leading to the accumulation of knowledge or skill: a lesson taught by experience

In the corporate world, most of the experiences are based in process driven events.
Scott Berkun says good procedures make people more efficient, not less efficient. This is because we need a guide on how to go about doing things, and it is more efficient to have processes in place, than having people doing things differently each time. Some of you might say or think about bad processes or bad data, yes, they do not contribute to any efficiency, but when we are willing to participate and share ideas, those processes automatically get fine tuned to a better level by the people involved reducing errors and enhancing accuracy.

One of the toughest experiences I had was while being in a hospital (I had an accident when I was a child, and another time I visited it with a depression issue), by observing ill people, doctors, nurses and the incredible job they do I learned an incredible amount I would not otherwise. I did not care or think much about my well being until that date, and have the experience of seeing other people suffer and deal with illness (My self included). That is the huge value experience shows us sometime like a bigger hammer, what works, what does not and what are the consequences of our own actions.

Something I found interesting from being in a hospital is that for many patients with depression, healing came by interacting and sharing their past experiences with other patients who where (intriguingly) connected to you in some form, same town, or different profession but same passion for what you do, and not exclusively in the doctor or drugs prescribed (or so I thought) . In my experience, people that had parallel lives or something in common made a big difference in the recovery and the possibility of recovering your health. Those interactions were incredibly helpful. The ones that did not have anything in common with you I felt had the opposite effect. They did not help or contribute to the recovery. I used to think doctors did not know what they were doing (how wrong I was); some are extremely excellent at what they do. Others have a very wide exposure to many people, allowing them to learn a lot from wider experiences.

In my conversations, I came to the conclusion that "Expertise" comes from ΣE or the sum of experiences that you go through. This is a good example that illustrates an area in live where experience is really crucial for people's health recovery.

Every process is a learning experience, and every experience is a learning process.

To summarize, the importance of experience and practice can be compiled in one sentence: "***An ounce of practice is worth more then tons of preaching.***" Gandhi

Information:

The simplest definition I found was: "Information is any represented pattern. This view assumes neither accuracy nor directly communicating parties, but instead assumes a separation between an object and its representation, as well as the involvement of someone capable of understanding this relationship. This view seems therefore to require a conscious mind" Reference.com June 23 2006.

I included this definition because information is only part of the rational side of the mind (Left), therefore requiring the left side to be "awake".

In acquiring knowledge about natural phenomena or event, the scientific method defines the process based on physical evidence and observation. Scientists use observations and measuring devices to gather data to propose new theories that explain the why of things. Deductions from these theories are then tested by experimentation. If a deduction or prediction turns out to be correct, the theory survives. A scientific method is essentially a cautious process of building supportable evidence-based understanding or our natural world.

The principal component of information is data. Then, when we apply a process to a data sample --Call it operations, or reflection -interpretation--- you get as a result information. Following the definition, information can only be defined as such after an individual makes an interpretation the data or a process using data. The insight is a representation of what the pattern the data is trying to communicate. Another aspect I disagree with is in putting information at the same level of knowledge. They do not sit at the same level as per our definition. Information is part of knowledge, but is not knowledge in itself because there experience is missing from it. To better understand, we can look at how decisions are made by business executives and doctors for example. They look at all the information available, opinions of people around. They try to gather as much input as possible from various sources. If information is incomplete or not accurate, the individual most of the times make the decision based on their "gut" feeling (See intuitive knowledge), but if not, they'll ponder information plus the widest experience available from other people and their own to reach the decision. .

Individual **Knowledge Model**

The interaction of elements generates information between data and intellect and processes. As an example In a big sheet of data, you bring it into Excel(Computer program), and then make operations like Sum, Avrg, or filter the information, sort it, etc. You get information (more insight) from applying a process (in this case math operation) to the data. All surveys take averages, maximums etc, to give you a conclusion about the data collected. On the other side having data and collaboration sitting in a room with a team of people and sharing it, can give you valuable information, but different from what you get by applying a process. See Linux and Toyota example below. By having data combined with collaboration you get more insight, with processes you will gain more accurate information.

The interaction between processes and collaboration is what I call experience. Many iterations of it bring expertise. An example: Let say you want to take the bike you learned to ride and want to go up the mountain. The process defines that you will be successful and gain experience only if you finish going from the bottom to the top. If you decide to stop 1/4 or half way, is not the same as if you complete the journey or ultimate goal, the top of the mountain. Another example is the process of buying a house. The goal is to purchase a property and have it under your name. Going through the steps but falling short at closing, does not mean you already "know" how to buy a house. Finishing the process is important in earning true experience, if you don not , then you only know parts of it. One byproduct of not completing the process is cost. In the bicycle example would be psychological cost, in the house example is monetary cost and probably both.

All elements interacting as a whole is what makes rational knowledge come true from different points of view (ways of thinking), opinions, and experiences.

There are three major areas in modern society that generate knowledge. Those are: Scientific Community, Marketing research companies and intelligence agencies. Two are prime examples of the model and one is a sad example of what happens when you have two out of three elements. First example: the scientific process or discovery. After developing a theory through observation or practice and after gathering data through trials, experiments, etc, the scientist creates a paper, or document detailing his (her) findings. The reason of publishing in a magazine is to have a wide audience so that ideas can be refuted or enhanced. I call it performing wider collaboration. You want people to collaborate with feedback on your ideas to see if there are holes or parts missing (as I did in this book), or if there is some unknown case where the theory does not apply.
The Internet is the by-product of scientists wanting to collaborate with a wider audience , making it even easier to accomplish it, and it is why Tim Berners Lee created the World Wide Web html scheme (and did not patent it or charged for it). The Internet is the biggest collaboration project in human history. The effects are rippling through every business and industry known, forcing many to reassess and rethink every business strategy from scratch.
The second example: Marketing companies like Gartner, IDC, and Forester rely on collaboration to bring insights to their clients. That insight comes in the form of surveys that when combined with statistical research, data and other elements provide extremely valuable insight or knowledge to someone. People have to be willing to fill out the surveys, in order to get meaningful and accurate data.

Some people do not see how collaboration plays a role in

acquiring knowledge. One way of testing the theory 'without collaboration you do not have knowledge' is by looking at what happened in 9/11. Simply put, a lot of intelligence agencies had a lot of information, but no sharing and collaboration was taking place. Hence you actually don't "know" what is going on in order to take Action. So if you have information , but do not share it and collaborate around, the outcome is simply you still have information, because many times the person responsible in doing something about it is not the same one holding the information in this case. Hence, knowledge (the rational side of it) requires collaboration between groups so that proper action can be taken. This translates into "the wider the audience, the more points of view you get, the closer to actual knowledge you will get". Another example I remember reading on the web, was about how the military use scientists to discover secrets on advance technologies (Flying devices) they have but that they did not create. In a very secretive environment, they expect a single scientist locked in a room to come out with all the knowledge on how these advanced technologies were built. My conclusion from it is that in a secretive setting, you can't go much further in understanding the unknown without collaborating with more people.

Unexpected Case Study for Individual knowledge

As the revised definition mentioned before, rational knowledge is about information and experience, we should be able to find evidence that proves it.

The Apprentice was a TV program in 2003 defined as a sixteen-week job interview, where eighteen people competed in a series of rigorous business tasks, many of which include well known companies and require information and experience to conquer, in order to show Donald Trump (the host/boss), that they are the best candidate for his company. In each episode, the losing team was sent to the boardroom, where Trump and his associates, judge the job applicants on their performance in the task. One person was fired and sent home each week.

If you missed it, was a very compelling program and very entertaining. I believe you can get a DVD with some additional footage. It is this very theory being put into practice.

When they decided to build two groups with the condition to have "Book smarts"(Those with Information through higher education) Vs Street smarts (Those with mostly Experience through work) I bet they did not know they would help us a lot!. It gave us a "experiment" to observe in real life how the two are related in normal people that were not acting up. Clearly people with more experience survived longer to the tests, but at the same time showed how women and their intuition steam rolled men at the beginning.

Two people lasted until the end, Bill and Kwame. Of the two,--clearly-- Bill was selected because he had a more balanced background (Education + Experience) against Kwame, and the results showed this equation. Bill had the book smarts (Information) and the street smarts (Experience) that made him the final winner.

Intuitive Knowledge

 In the book "Awakening Intuition", MD. Mona Lisa S writes: "Intuition is the process of reaching accurate conclusions based on inaccurate and many times incomplete information"., she talks about intuition being like a radio station, always on, sending us messages through different channels (the gut as one of them), and the way the mind and the body are connected; When we do not tune in to the radio channels, it is very likely that a disease will appear as a result of not heading to the message by not paying attention. Another way to listen "the intuition radio station" is by remembering our dreams.

> Moses Maimonides said "Tell me what your dreams are, and I will tell you not only what you are, but what you are to become."

> "A night dream comes with a purpose of aligning us with the present moment and showing us- to us. When looking into a dream we are looking into a mirror. In our waking life when we look into a mirror we see quantities of ourselves; that is; one nose, two eyes… In the dream life we look into qualities of ourselves represented by characters in the dream. Any person, place or event in the dream holds tremendous significance for understanding ourselves. Nightmares are simply messages from the deepest part of us to our consciousness calling for change. If unheeded, not only may we continue to suffer from the unpleasantness of a "bad" dream, but we run the risk of perpetuating negativity in our waking life. And as MS states, in will shift to illness

to force change. Rudiger Dahlke, in his book "**The Healing Power of Illness"** explains that contrary to conventional opinion, illness is not some quirk of nature you have to fight. A truer understanding of illness actually helps you stay healthier and make changes sooner rather than later. When you "understand what your symptoms are telling you," you can view them as bodily expressions of inner conflict and your body is the intuitive messenger.

By uncovering the language and symbolism of dreams we can learn about:

1. Our relationship with the world and ourselves around the time of the dream.
2. The "global" issues (physical and emotional challenges) that we face in our lives.
3. The condition of our body at the time of the dream.
4. Our unconscious beliefs.
5. How to solve our problems." By Dr.Peter Reznik [15]

When Deep Blue's computer power defeated worlds chess champion Gary Gasparov in 1997, it send shock waves to the western world. In the east, news about a computer wining a chess champion did not impress much. Why? "While there are avid chess players in Japan, China, Korea and throughout the East, far more popular is the deceptively simple game of Go, in which black and white pieces called stones are used to form intricate, interlocking patterns that sprawl across the board. Go fans proudly note, a computer has not come close to mastering what remains a uniquely human game. To play a decent game of Go, a computer must be endowed with the ability to recognize subtle, complex patterns and to draw on the kind of intuitive knowledge needed that is the hallmark of human intelligence. Something not easy to replicate in a software algorithm. Expert Go players evaluate the state of the board by using their skills at pattern recognition, and these are very hard to capture in a computer program. After years of experience, they can look at a complex configurations and sense whether it is "alive," meaning that it is constructed in such a way that it cannot be captured, or "dead," so that no amount of reinforcement can save it. Learning to sense life and death is the key. An example on how rational vs intuitive knowledge plays in a game environment. It shows that as powerful as they might look, machines might never equal the power of the human brain.

Intuition has always played a key role in corporate settings, although not many may realize it.

[15] @ http://www.drpeterreznik.com/articles_dreamwork.html

"Roger Saillant, CEO of Plug Power, one of the first electric fuel cell companies, has good instincts. Saillant has created an organization that *feels* different, that has an energy that is palpable. Work at Plug Power "is not *your* job or *my* job. It's *our* job," he states. "And that's how people become enlisted when we are working together. It is what happens when you think of yourself as having no boundaries, when you think of yourself as working in a field of connection and consciousness." In creating this organization, Saillant has tapped into something that moves human beings and not just machines: "I believe that people want the truth; they want to learn and grow, to be part of a community and a shared inspirational vision," he states. "When you try to practice these principles, somehow the universe reaches out and gives you insights that guide you at an **intuitive** level." No longer isolated in the command tower, Saillant is part of a neural network of human relationship that learns and grows together" From WIE.org

In my personal case, I learned little use of my intuition, until an illness hit me, it made me aware how unbalanced my mind was. From there I learned to play piano, and started writing with my left hand. After some valuable guidance from a good friend of mine, I started watching more closely the dreams I had.

Summary:

1- There are two types (Forms) of knowledge, based on our brain physiology.

Rational knowledge - Left hemisphere

Intuitive knowledge. - Right hemisphere.

Our goal is to look deeper into rational knowledge concept and how it gets generated vs Intuitive knowledge which is a topic for another book.

Case study Intuitive knowledge

Excerpt From Guardian February 2005.
Daniel Tammet is an autistic savant[16]. He can perform mind-boggling mathematical calculations at breakneck speeds. But

---

[16] Savant definition: From French word savant meaning Scientist, literally knowing. Person who expresses extraordinary mental abilities, It is sometimes acquired in an accident or illness, typically one that impairs the left side of the brain. There is evidence that suggest savant abilities are innate within all of us but obscured by the normal functioning intellect (Left hemisphere)

unlike other savants, who can perform similar feats, Tammet can describe how he does it. He speaks seven languages and is even devising his own language. Since his epileptic fit, he has been able to see numbers as shapes, colors and textures. The number two, for instance, is a motion, and five is a clap of thunder. "When I multiply numbers together, I see two shapes. The image starts to change and evolve, and a third shape emerges. That's the answer. It's mental imagery. It's like math without having to think." Now he is 27, and a mathematical genius who can figure out cube roots quicker than a calculator and recall pi to 22,514 decimal places. He found it easy, he says, because he didn't even have to "think". To him, pi isn't an abstract set of digits; it's a visual story, a film projected in front of his eyes. He learnt the number forwards and backwards and, last year, spent five hours recalling it in front of an adjudicator. He wanted to prove a point. "I memorized pi to 22,514 decimal places, and I am technically disabled. I just wanted to show people that disability needn't get in the way." Famous autistic savants have right brain qualities, like number calculator, music and arts, and incredible language skills and all because their left side brain stops, and the right side is the one working.

Personal knowledge Vs Collective knowledge

In this section I go deeper into what the relation is between individual knowledge concept and collective knowledge.

We defined individual knowledge = information + experience, which is what an individual is able to learn on his own since birth. I will give it the name iK; when you combine collaboration with personal knowledge I call it cK [17], bringing together multiple experiences and information repositories, something known in general as the wisdom of crowds[17]. An expanded view of many people experiences and points of view, produce a Social Network which is a social structure made of nodes which are generally individuals or organizations. Research in a number of academic fields have demonstrated that social networks operate on many levels, from families up to the level of nation, and play a critical role in determining the way problems are solved, organizations are run, and the degree to which individuals succeed in achieving their goals.

We can then define cK= individual knowledge + Collaboration. Knowledge is represented in the element interactions –darker colors, following a system definition. If you take an elephant –a system-- and divided it in two, you do not have two elephants as a result. You get a mess of parts as the integrity of something called

---

[17] Wisdom of Crowds term defined by James Surowiecki

elephant was broken. Data only or processes by themselves do not give you collective knowledge, only the interaction between all will. Collaboration the goal we have been missing to make KM a successful practice.

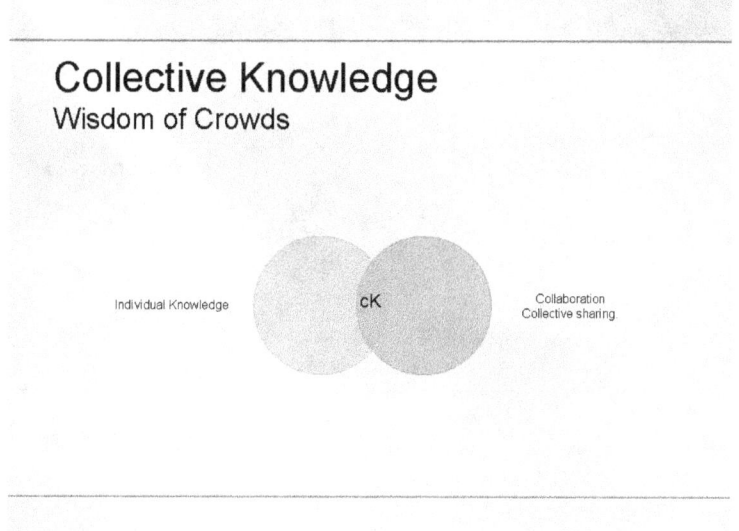

Figure 4 Collective Knowledge

There are additional elements to consider when interacting with knowledge from the technology perspective. These are:

Interface: it represents the medium and the form that we are using to access the systems information. In the past it was a book as the medium to communicate, today it means it could be a computer, electronic device, cell phone, etc used to access information. The key aspect of it is its flexibility and adaptability to any situation, so that people will always be able to read the information presented.

Find: I like more this word than "searching" because you can search forever and never find anything meaningful. Finding means just that, been successful at searching something, sorting it out, and extracting what you are looking for. Google makes a great job at bringing us relevant results based on the value it calculates from the connections or hyperlinks in pages around the web. That is a good first step into what the future could bring easy tools that allow us to scan and FIND the entire information universe.

In summary, the efficient storage of data and information combined with other elements (Experience, collaboration and technology) constitutes the base and foundation for human rational knowledge today.

Personal *Wisdom*

"Wisdom is the supreme part of happiness." Sophocles:

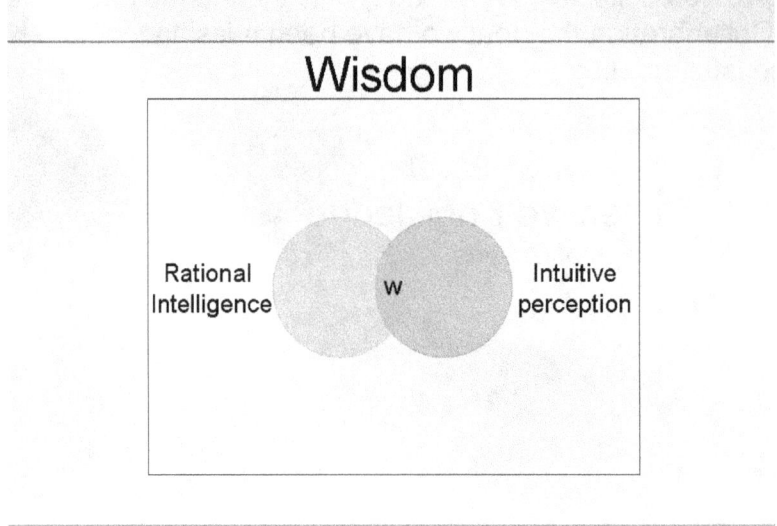

Figure 5 Wisdom diagram

In the dictionary we find that wisdom is "accumulated knowledge or erudition or enlightenment 2: the trait of utilizing knowledge and experience with common sense and insight. Personal Wisdom is the balance by using the two sides of our brain.
For me it is the trait of utilizing reason, experience, and intuition with common sense and insight.

According to Wikipedia, the most famous examples of wisdom literature are found in the Bible. The following Biblical books are classified as wisdom literature:

- *Psalms*
- *Proverbs*
- *Ecclesiastes*
- *Song of Songs*
- *Wisdom*
- *Sirach*

"Even as wisdom often comes from the mouths of babies, so does it often come from the mouths of old people. The golden rule is to test everything in the light of reason and experience, no matter from where it comes." Mahatma Ghandi

Case Studies on Collective knowledge

Wikipedia

A radical new idea created by Jimmy Whales, it has proven to be a highly popular web site. The reason: Creating a repository free for all, where anyone can see, edit, add articles has transformed Wikipedia into a colossus collaborative product. One million plus articles in English language makes it twelve times the size of Encyclopedia Britannica. The other two million articles it has are under 200 other languages. Over one hundred thousand people have contributed to it around the world as of June 2006. Wiki comes from the Hawaiian word for "quick", but it also stands for "What I know is… ", they are thus the simplest form of collective knowledge sharing or "wisdom of crowds" as reviewed earlier. As any other product, it is very new with five years only, which means is not perfect, and has a lot of criticism on the process itself for publishing information and it's quality. Some people believe articles are not accurate, and the process to measure quality in a comparable product is questionable, that is "the quality depends on the total number of bad articles", in my mind I would say is the opposite, if the number of good articles far surpasses the bad, then it is a good resource, but measures should be put in place to quickly solve disputes for bad inputs. And because as I write in this book, knowledge comes from information and experience, I experienced my self Wikipedia by trying to modify their explanation of the word knowledge. Overall it was not very successful, because it is not available as a regular entry, but converted into an epistemology project (Philosophy), where a filtering system of few people rejecting entries worked like this until I stopped doing it: If your entry is not "encyclopedic" then it gets rejected. Maybe because I come from IT with a different way of viewing knowledge, it is not "encyclopedic", here is it's definition (A written compendium of knowledge, from Greek "A well rounded education") so the reason I was been given has no sense, but I was starting this book, and had no plans to get into fighting mode. After some time I thought about what could be done to solve this issue. One solution might be to let people post "versions" which they already have, but have others vote for the ones they understand the most. The most popular, get to the top of the pile. I imagine having the whole spectrum visible, from the worst article, to the best. That would not only give us insights about what is the best definition, but also the opposite. Something along these lines has been already created. It is called Digg.

Digg
A news website created by Kevin Rose where complete control is given to the community, a site where users could submit stories that would fall into a general queue, and if they were popular enough--if they got enough 'diggs' [user endorsements]--they

would be promoted to the homepage for everyone to see." he says. In less than 2 years, his service climbed up from 12.000 users to more than 200.000 today serving 7 million pages a day. In my research, I find amusing people objecting to this idea, by saying "crowds aren't all that wise", or "personalized news are bad for democracy". I thought democracy meant "rule by the people". It shows you will just find any sort of comments no matter what you create; it helps to remind ourselves to disregard negative comments as fast as they appear.

Generating Knowledge

How do we learn rationally?

Learning definition; It is the process of acquiring knowledge (Information for our purpose) or skills through study, experience or teaching.

How do we Learn? Here is a list in the order we may learn since we are born.

- Through Imitation

I learned the importance of it while being depressed. The depressed mental state people are more likely to recover by simply imitating the actions of somebody else. Which actions? Actions that reflect discipline, order, cleanliness, and basic social skills like talking to others. Babies and toddlers learn their first skills through imitation. The first thing my daughter learned was smiling and clapping her hands though imitation.

- Through Experimentation/Practice/Experience

The next stage in babies development is through practicing, and experimenting with objects, colors, food and tastes, you teach them how to sit and hold steady, and then how to roll over from an upside position. Little by little they learn all the little tricks that we take for granted. For them, it takes time and practice.

- Through Teaching

My best example definitely comes from potty training a young toddler. We were sent by family members a book (toilet training in less than a day by Dr, Nathan Azrin , Richard Foxx). I thought it was a very interesting technique to try but awfully difficult to remember all the steps meaning all the preparation required. I felt like going back to school. The result: Just amazing; our daughter at the age of two and three months was guided by me and her mother into teaching her doll, how to go to the potty first. So by having her teach the doll, the theory was she would understand better the concept. After studying all the steps and buying all the things needed we chose a long weekend to give it a try. We thought that in two or three days we could be done. We told Isabella to guide the doll into lowering her pants, feeling them and recognizing "dry" from "wet", and the importance to keep them dry, because she was not going to use anymore diapers, then watch the doll make pipi, and then have her clean herself (The doll), raise her pants, take the seat out of the potty, the potty out and throw the pipi into the toilet, and flush the toilet. Then we told Isabella that it was her turn to be a big girl, and followed all the basic instructions she gave her doll and others in the book. I added a couple more steps like taking paper to clean the potty, and washing her hands on her own. My goal was to see if she could do the complete set of tasks, so that she would be totally independent, and leave the bath room the same way she found it. The book did not include all steps I added, but I was so interested in learning by teaching technique that I wanted to see how much she could learn. She learnt the basics in little over half a day, having the usual "accidents" and our reinforcement practice after them. In the night we decided to use pull ups, until she learnt how to contain herself "Unconsciously". We thought it would take a couple of months. Big surprise when in day 5, she woke up 3 AM to call for pipi. The following nights her pull up pants became drier and drier. Something I realized was I unconsciously gave her double instructions (lack of patience), so with this exercise I had to force my self to slow down and say things once. After a week or so, I decided to stop talking and just use signs to see if she would respond faster. I was shocked when Isabella started to respond immediately to certain commands by hand gestures. To make her stop doing something, I would say "stop". The second week, she had an accident while taking a nap in the middle of the day, that caught us off. Her reaction when she woke up made us understand how important for her was. Tears were filling her eyes at realizing what she did. We comforted her, and told her she would sleep in the lower bed, while the bigger would dry. Next night, the bed was dry, but even when I told her it was ok to sleep in her big bed, she refused, saying "pipi dry". This has been an incredible learning experience for us.

- Through Observation, Environment, situation.

Most of what the Scientific method teaches, and scientists do is through this method. Nature's researchers use observation to learn behaviors, skills, etc from animals, insects, plants and in general of Mother Nature's wisdom.

- Through Study (Reading, memorizing)

Once humans learn the skill to read and write, we can gain insight through study of books and other material about various aspects of life, science, religion, etc.

- By Touching, Sensing (Experience)

One of the five senses, touching allows us to learn the difference between soft, hard, rough, smooth, forms like round, square (When not observing) and many others.

- By Listening (Experience)

By listening I mean active listening focusing our mind in the process, which is different from listening but not paying attention. It is probably one of the skills I have to work constantly more than others. The importance of separating all the senses is in the fact that the mind has to be focused in that particular sense in order to gain full perception.

- By Smelling (Experience)

Self explanatory

- Through repetition (Experience, Memorizing).

We learn many times by simply repeating a task, either by memorizing a text or image, or by doing over and over again an action. This is also part of experience and practice.

- By reflecting. (Reasoning, contemplation)

We can gain additional insight and understanding from experiences or study, by reflecting about them. It happens to me often , when having a debate through Instant Messenger I write certain things that later that day when reflecting about what it was wrote, I realize there was a better way to say it, or more information I could have added to make it more complete. We often do this with past experiences in a negative way, meaning we start worrying about an event, instead of looking for lessons and moving on.

The Amazing Jelly Bean Experiment

Treynor asked his class to estimate how many jelly beans there were in a jar. When added together and averaged, the group's estimate was 871— there were 850 beans contained within the jar. Only one student had made a better guess (a rogue genius, if you will). The now historic jelly-beans-in-the-jar experiment showed invariably that a group estimate average is superior or more accurate to the vast majority of individual guesses on a consistent basis.

Granted, there are limited situations in which knowing the amount of jelly beans in a jar is a significant accomplishment. Nevertheless, this example can be found a book by James Surowiecki called *The Wisdom of Crowds*.

"If four basic conditions are met, a crowd's 'collective intelligence' will produce better outcomes than a small group of experts, Surowiecki says, even if members of the crowd don't know all the facts or choose, individually, to act irrationally. 'Wise crowds' need
(1) Diversity of opinion;
 (2) Independence of members from one another;
(3) Decentralization; and
(4) A good method for aggregating opinions." —*Publisher's Weekly*

To validate knowledge in all situations and make bigger contributions to the whole, we need to share it with many others and aggregate their feedback because they will attest through their own experience (and not only logic) that the information you present is true or need some adjustment; in the diagram below only when you combine information + experience + collaboration or K=D+P+C, you'll get validation and better comprehension of a given theory and practice. In other words, you can learn something new through I+E, but you'll find that there are situations where either it does not apply, or there are other conditions that you as a group need to take into account. HBS professor Kent Bowen, a technology and operations management expert who has studied Toyota, explains: "The Japanese are very good at two things that are key to success in the auto industry: refreshing their products, and having the flexibility in their factories to do that quickly and economically. Toyota "strives to use its labor force in flexible, creative, and collaborative ways," Bowen says. Snow adds that in their factories, the Japanese often prefer to customize existing systems and equipment rather than installing the latest fancy technology. "Their plants are in a state of continual improvement and repurposing, **with input from everybody**," Snow notes.
In developing a new device or product, only when many people use it under different circumstances you'll learn how to make it

better (if you incorporate their feedback back into the product) as Toyota does. Even if your team is talented, they won't be able to come up with every possible situation where the device/product might fail. Knowledge is only useful to others in the degree you share (it) and collaborate. The more you collaborate, receive feedback and incorporate others ideas, the more wisdom you'll have.

Last example I want to mention about collaboration is Linux and Toyota. The article link can be found here

Linux and Toyota, the power of collaboration

Extracted from "Collaboration Rules," **Harvard Business Review**, Vol. 83, No. 7, July/August 2005.

"Tuesday, December 2, 2003

Near midnight, Andrea Barisani, system administrator in the physics department of the University of Trieste, discovered that an attacker had struck his institution's Gentoo Linux server. He traced the breach to a vulnerable spot in the Linux kernel and another in rsync, a file transfer mechanism that automatically replicates data among computers. This was a serious attack: Any penetration of rsync could compromise files in thousands of servers worldwide. Barisani woke some colleagues, who put him in touch with Mike Warfield, a senior researcher at Internet Security Systems in Atlanta, and with Andrew "Tridge" Tridgell, a well-known Linux programmer in Australia on whose doctoral thesis rsync was based. They directed Barisani's message (made anonymous for security reasons) to another Australian, Martin Pool, who worked for Hewlett-Packard in Canberra and had been a leader in rsync's development. Although Pool was no longer responsible for rsync (nobody was), he immediately hit the phones and e-mail, first quizzing Warfield and Dave Dykstra (another early contributor to rsync's development, who was based in California) about vulnerabilities and then helping Barisani trace the failure line by line.

By morning Trieste time, Pool and Barisani had found the precise location of the breach. Pool contacted the current rsync development group, while Barisani connected with the loose affiliation of amateurs and professionals that package Gentoo Linux, and he posted an early warning advisory to the Gentoo site. Pool and Paul "Rusty" Russell (a fellow Canberran who works for IBM) then labored through the Australian night to write a patch, and within five hours Gentoo user-developers started testing the first version. Meanwhile, Tridge crafted a description of the vulnerability and its fix, being sure (at Pool's urging) to credit Barisani and Warfield for their behind-the-scenes efforts. On Thursday afternoon Canberra time, the announcement and the patch were posted to the rsync Web site and thus distributed to Linux users worldwide.

Common intellectual property. The General Public License under which Linux is published requires that all distributors make their source code freely available so that others can freely emend it. This viral principle prevents code from being stowed away in proprietary products. That transparency, in turn, breaks down the distinction between producer and user. A sophisticated "customer" like Andrea Barisani is really a user-developer, who fixes flaws and adds features for his own benefit, then shares those improvements with everyone else. Such a role is impossible when proprietary code is licensed from a commercial vendor. Similarly, Toyota's supply chain is predicated on the principle that while product knowledge (such as a blueprint) is someone's intellectual property, process knowledge is shared. That breaks down some distinctions among companies. Toyota's suppliers regularly share extensive process improvement lessons both vertically and laterally, even with their competitors. In Japan, suppliers are generally exclusive to a single OEM, so the collective benefit of that shared information stays within the Toyota supply chain. But even in the United States, where Toyota is just one of several customers for most of its tier ones, the carmaker does the same thing through its Bluegrass Automotive Manufacturers Association, which disseminates best practices to all members.

Simple, pervasive technology. Although information is the lifeblood of the Linux and TPS communities, their circulation systems are surprisingly rudimentary. Linux developers produce state-of-the-art software using communication technology no more sophisticated than e-mail and Listservs—but those mundane tools are used by everyone. Indeed, so great is the value placed on universality that plain-text, rather than formatted, e-mails are the norm, ensuring that messages will appear exactly the same to all recipients. Toyota, whose products are state-of-the-art as well, also prefers simple and pervasive internal technology. An empty kanban bin signals the need for parts replenishment; a length of duct tape on the assembly-line floor allots the completion times of tasks on a moving vehicle. Quality control problems on the assembly line are announced via pagers and TV monitors. And everyone gets the alert. Even Ray Tanguay, head of Toyota Canada, is paged whenever a flaw is found in the latest Lexus consignment on the dock in Long Beach, California, or in a service bay anywhere in North America. [...]
Such extremely rich, flexible collaborations have positive psychological consequences for participants and powerful competitive ones for their organizations. Those consequences are

rich common knowledge, the ability to organize teams modularly, extraordinary motivation, and high levels of trust.
" From "Collaboration Rules," **Harvard Business Review**, Vol. 83, No. 7, July/August 2005.
The Internet is proving what collaboration is capable of. As of today (May 2006) I can find many examples where the Internet is breaking every paradigm and industry upside down. Real state, publishing, Sales, marketing, just name it, and you'll see a tidal wave coming if not already happening. This is the effect of human collaboration at great scale.

With these elements clear, from the IT perspective, we can transform the simple concept of knowledge into a powerful software application that might transform the way we think and work today and tomorrow (See chapter 4).

The solution is in the designed solution to transform these knowledge concepts into software tools that provide a platform and the means for anyone to build their own knowledge base on top of it. We can translate the knowledge requirements into a simple, yet powerful software platform that can provide the means for achieving 100% collaboration. Before we dive into technology, we need to look at the people side. Next chapter will introduce you to the Learning Organization. I believe people always go first, then technology.

New Leadership

Leadership Wisdom

"Two qualities for leadership are to be a visionary and to know execution," says Dr. V[18]. "If I can go from consciousness to higher consciousness, then I'll be a leader." "How can my work make me a better human being and make a better world?"
To achieve perfection, it helps to respect money -- but not to be motivated by it.
The reward for work is not what you get out of it but what you become from it.
Dr. V. considers his gifts to be the things that he has given others, not what he possesses.
Give people a new experience, one that deeply changes their lives, make it affordable, and eventually you change the whole world. And your customers become your marketers.
Dr. V. teaches that work can be a vehicle for self-transcendence.
"If You Are Looking for a Big Opportunity, Find a Big Problem."

 Your ability to succeed depends in great measure on your capacity to learn, plan and collaborate ideas forming a broader knowledge base for building better solutions.

Many organizations believe that moving quickly will always keep you them at the front, but based on "systems thinking" theories, faster, quicker and pushing excessive growth in a system produces the system to compensate by slowing down (The more you push, the more it pushes back). The rise and fall of many companies have many times demonstrated this common phenomenon. People Express airline story in the 1980's, tell us about how an airline with superb customer service got to be one of the best and most profitable very quickly, but after 6 years, it came down rapidly to a halt and into bankruptcy. The high pace growing up was not compensated by an increase in labor training on customer service. Therefore the more they grew, the worst customer service turned to be, the less people continued to fly. Instead of increasing prices to control growth and fund training

---

[18] *"Dr. V(enkataswamy), the legendary eye surgeon from South India, who with his own two hands restored the sight of over 100,000 people. His work resulted in one of the world's most extraordinary models of service delivery. Thirty years ago, at the age of 58, he started an 11-bed eye clinic in an old temple-city, and with his team, turned Aravind Eye Care System into the largest and most productive eye care facility in the world." From SOL. http://www.signsoflight.org/more.php?n=2472*

service associates, the company focused on more sales. It's demise could be seen long before it came to reality.

Collaboration is only fully effective in a "Learning Organization". While people collaborate, they learn new insights as to how things should be done, and especially making sure ideas have long and lasting results.

Decision making.

One day, I glanced a book title that said: "I prefer to act and make mistakes, than doubt and do nothing". It should be allowed in society and corporations to make mistakes, but western world punish since early education stages, those who make mistakes. How can we learn if we are not allowed to make mistakes? . Making mistakes does not mean you can make the same mistake over and over again. That is foolish. The goal of a mistake is to teach you what does not work, for you to then correct and come up with the solution. Not willing to learn from mistakes is something I can't comprehend. We can't find solutions without learning and knowing that something is failing.

Gary Klein observed in his book How people make decisions (MIT Press,1999) that project managers rarely have enough information and time to make decisions (Using methodologies) work. They have four things: experience, intuition, training and each other. He thought they make good decisions by maximizing those elements. He mentions the two types of knowledge --rational and intuitive, and experience/training or education plus each other (Collaboration) as the way you can make decisions based on all knowledge available.

Decisions are a good measure to observe at how we use the different types of knowledge to take decisive action to solve problems. Inexperienced individuals may suffer by only relying on information glanced from say the internet to drive major decisions. I witnessed it many times. Decisions made based only on reports from a research company, who conducted a study, without taking into consideration people around the decision maker as to which was the correct path, when many had the information, and raised red flags. Decisions made only on numbers like cost, and not on actual business needs have the biggest negative impact in the long term for a business. Cost cutting without looking at user impact is counter productive.

Different set of leadership priorities

Example: Google

"Companies need to consider opening their flow of information", said Google Enterprise General Manager Dave Girouard

Product management has evolved from being formulaic, linear, scheduled and predictable, to being unpredictable, cyclical and revolutionary. "You can't schedule innovation," Girouard said. Google is making big noise on the products they are developing, and they are changing the "corporate paradigm" upside down by being open.

"Companies need to attract, guide and retain self-directed innovators in order to succeed in the next 10 to 20 years. --Self-directed innovators-- who don't necessarily hold high-ranking positions have a wide influence within a company and can make things happen". "Those employees can't always recall details of all of the information they have come across so they need a photographic memory, collective wisdom and the world's information, he said. The way to deal with that, according to Girouard, is to allow information to flow more freely and quickly than it has in the past" he said. He outlined the top five ways to maintain an outdated business model. They are: restricting internal publishing by failing to provide a physical means to communicate or by instituting strict rules about gaining approval; requiring publishers to follow strict metadata standards; assuming employees will use systems because they're in place rather than letting employees' behavior give cues about what works, building overly complex interfaces, and restricting access when in doubt.

Successful thinking model

What does success looks like?

It was odd to say the least to hear the question above because we assume the answer is clear. We think is logical. Find the problem, then solve it. Time for another trip to the dictionary. Success is "The achievement of something desired, planned, and executed". As nobody responded in the weeks after, I kept
thinking if there was an effective way to demonstrate how it looks like. Maybe a visual slide show might do the trick. I decided to give

it a try, but where to start? After a while, an idea came suggesting trying the opposite. Success is the *opposite* of failure (see thesaurus). I wrote it down. Then I reflected on other meeting discussions with technology and started to draw a map not knowing where it would lead.

In Figure 6 What does success looks like, I typed technology and went back to the dictionary to check the meaning (Technology comes from techne (τεχνη) "craft" + logia (λογια) "saying") meaning the science about tools created by man applied in conjunction to build a solution to a defined problem. Tools were not the solution itself, but the way you can create your vision. *"Technology is a tool that can change the nature of learning"* Lynne Schrum.

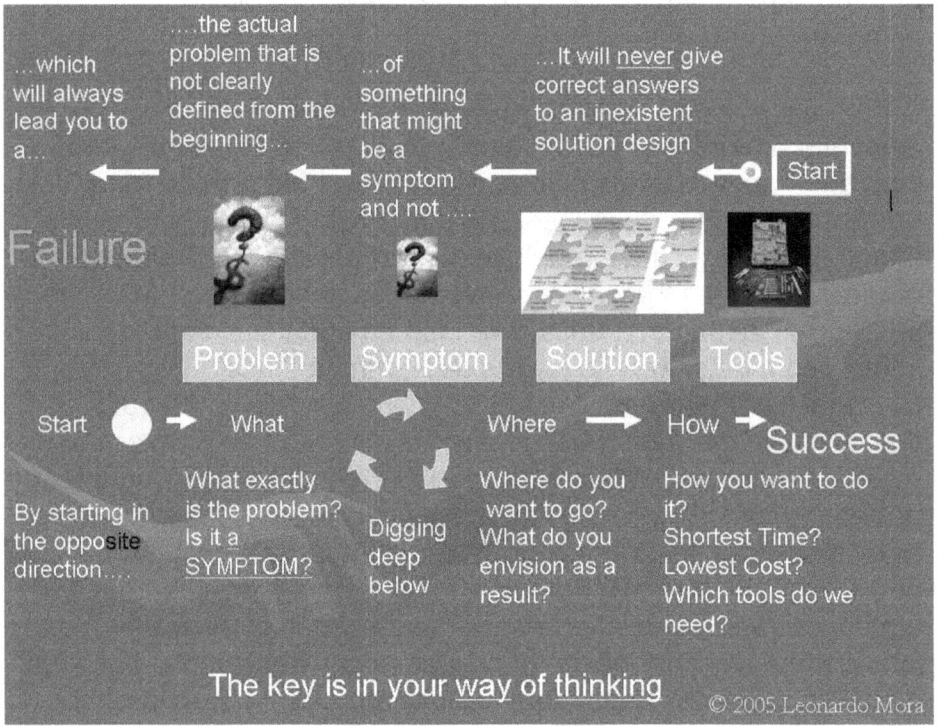

By observing people around in the office, I learned that they wanted answers from the tool, that would give the solution clearly and directly, and then members of the team would look at the

Figure 6 What does success looks like

requirements (which define the problem) after the fact (Choosing the software tool). I kept staring at the slide for a moment in disbelief. At the end of that path (Right to left), I wrote"Failure". Why? I could not find a method, where tools (or technology) can give you all the right answers to a non existent solution (That it has not been envisioned) of a problem that has not been defined and agreed from the very beginning (See Diagram); Even more grave, it means you might be trying to solve a symptom, not the core problem as we see today all around. In one case, a consulting company was migrating information in a certain project; the client let them know that he wanted one million documents inside a library container (specific software feature) in their application - solution: the developer amazed by the request replied that the container feature in the application was not designed for holding one million records. It never crossed the developer's mind that such a requirement might exist. The obvious response from a client or customer in situations like this is naturally "That is what I need, make it work" or some type of angry response. These kind of "crazy" requests that I see all the time are the consequence of people choosing a tool first, and then trying to cram a need into a software feature not designed for it. In another case, a team was trying to group all types of documents (Contracts, memos, letters) in one container as the way to do it based on the software functionality. Although it might look helpful, the real world use of it might be different, as many of those files are used in different groups or units. Managers might benefit from the grouping method having broad views of information, but content creators and users within units do not have the same perspective. They want information grouped based on their unit, not across all of them. Clearly the solution is a mix between the two but until people like you and me realize that we need to design and envision solutions first, we'll continue hearing crazy "shove" software stories. The fundamental problem and solution should be formulated in a non technological language.

By solution, I am not referring to what you built. The actual solution is what you imagined it would work as the solution to the problem, the design. The airplane built is not the solution, is just the reflection that your ideas and design might work. The FAA (Federal Aviation Administration) will not give your plane a license to operate; until you fly it and prove to them it works properly and safely. The same happens in construction. A COO (Certificate of occupancy) is only issued once an inspector verifies that what you built is sound and safe for people to live inside.

The Problem, and the envisioned solution must be simple and clearly written down for everybody and anybody to understand. Then and only then, you along with technical people can choose which tools might fit the specs for building it.

Example: The company employees take too much time searching and retrieving information. This is a symptom. The heart of problem relies on how is the information organized and stored. One solution is to have a big bag where everything goes in, and then you search for it (librarian style) See figure 2

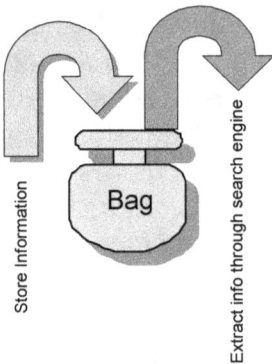

Store Information

Extract info through search engine

Bag

Figure 7 Information bag

, the other is to create a structure and views of the different ways people access information (Solution). How we do it is left to the tool or tools. The issue with "Bag" thinking is information does not relate to each other, leaving us with only one way of looking for it, the search engine, that in many cases is not accurate.

In another case, I observed Kathy trying to link or map a problem directly to a tool (Software) thinking that without a solution, the tool could actually be linked to the problem.
Going in the opposite direction (as in the diagram right to left) chances are you will find success by clearly defining what the core problem is separating symptoms, what the solution should be and should look like, and then start defining HOW you will build it with the tools and technology. Technical language should only reside in a technical specification. In some cases like when the mouse was invented, the tool might not exist, requiring a new material or tool to be created, in the mouse, the rolling ball material was not adequate, because it would gather so much dirt, it would make it impossible to use.

I call this method the **PSST system** concept (Problem ->Symptom->Solution -> Tools)

I sent an email with the presentation to the person who requested it. Certainly, a quick way to get in trouble (so I feared), but I decided to do it no matter what after getting some inspiration. That's what a consultant should do anyway, right? To tell (the customer) what is wrong and the way people should solve problems. The answer was what surprised me the most. None.

Let's look at the line of thought from another perspective -- construction.

"What is the *very* first thing you need to do when you want to build?",

0. Should I build? Decide if I should build or not
1. Your selection will depend on your needs and resources, and they should be based on your vision. If they are very specific (Which in knowledge and business terms they always are) you will have to be careful and create a detailed list of them. When your needs are unique, most likely what is already there in the housing market may not fit your current and future requirements. If they do, you might find trouble in the long term as you grow your needs and family, requiring either customizing your headquarters to the point where you reach the limits of what you can do, having to move again, or you might decide that building a structure designed and planned for the long run is the best choice. Finding out why you need to build will always be the key to keep you going.
2. Next you need to look at which resources you need to accomplish the task (Creativity is crucial here); you need to define the form and type, is it a house, apartment, hut? The decision will come normally based on your income or buying power and we'll assume you are not in a deserted island. If you can not build, because you have no resources (Money, people) then you will need to rent, so building a new place is out of the question. Now, in the corporate world, this translates to either continue with what you have, or build something new. The danger is relying on a vendor "solution" as we said for implementing what you need.
3. Assuming you have the resources is common wisdom that you first hire an architect to design the house or structure you need. Normally you want to see it before you start working in it, because you want to be able to make changes once you see how it will look like. Those changes are easier and far less expensive to do in a design stage, than in the execution stage. Why? Any designer will tell you that the elements in a design are not just a list of parts, but a whole entity. This means that if you make a change, the designer will have to rethink all the structure to maintain its integrity (keep other requirements functional). This is by no means easy work. If you make many changes, most likely you will need to start from scratch, and redefine your needs because there is a point where a couple of changes affect many other aspects or requirements, making your solution "unstable". Take an airplane. If you make too many changes (like bigger wings, bigger engines or bigger body), you are better off designing a new airplane than trying to modify an existing one. The risks of failure are too great when changing the wing span, or enlarging the main body to accommodate more passengers. If you hire an engineer

first then the engineer's role is to build you a house, not to envision or design it. If they provide you with a design, then he is the architect as well. Strictly talking, an Engineer role is to take "Blue prints" in order to build your solution. In the IT world it should be the same, but it does not happen in this order. In fact is the opposite.

4. Once you have the design, made the changes, then you are ready to start building. (I assume you own the plot of land) Once in the construction phase, few major changes are allowed. In IT is no different. If the project manager allows big changes to happen, the project will likely fail.

5. Hopefully, in couple months, engineers and workers will finish your house. Adjustments are likely to happen, but the house will look much like the design the customer approved. It is therefore a successful project.

We just reviewed one aspect of the problem. The second challenge is actually moving from A to B (your current situation to the future situation. In information terms, it is in many cases a whole project on its own given the scale of companies and the rate at which information grows year per year.

The third obstacle to KM comes with dealing with the term "sharing knowledge" which few people find to be the fundamental goal. Sharing knowledge will always involve people, and people will not share or provide *all* their knowledge easily just because you ask them or try to enforce it with compliance and policies. It involves a lot more than just technology and compliance. Sharing knowledge means convincing and leading people that it is beneficial for them and the company in conjunction. How? I found that the only effective way (As mentioned earlier) is to completely change the corporate culture. For more information, please jump to Learning Organization section.

The main effect of not paying attention to failure is that you are bound to repeat over and over the same practice without learning the lessons, extending failure indefinitely. Once you are in that cycle, you become blind/deaf to warning messages trying to tell you what your fundamental problem is. Sending red flag without response has been my experience.

Tools & Technology

It is important to distinguish what are tools, and what is a solution. In the above example, the house design was the solution to be built, and the tools were the devices used to build it.

Figure 8 Tools explanation

Hardware refers to the computer or device, a tool by definition. (See Figure 4). By Technology, I am referring to hardware and software. Software can be divided into two general levels or classes: Systems software and applications software.

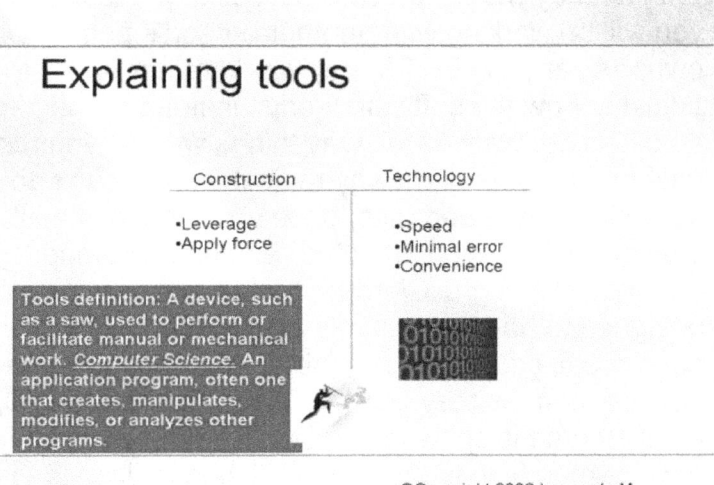

©Copyright 2006 Leonardo Mora

Systems software is the low level programs (s) that interact with the computer at the most basic level. It includes operating systems, compilers, etc. For our purpose, they are clearly underlying tools that enable other applications to run properly.

On the other hand, application software is one or several programs designed for end users through development applications, or languages like Basic, C++, Perl, etc. These languages allow you to build many different kinds of programs for end users benefit. I consider them developing and developer tools. For instance, Microsoft word is a program designed to create written documents. Applications should always start from an algorithm and interface design, the basis of a solution. The program itself is the result. The house built. So we can create solutions with software, but as in construction, the program will answer specific needs laid in a design document. We should not claim that a given house that is already built solve all housing problems. It solved a particular set of needs for the owner who built it. Other individuals would have to make sure that their needs fit into the existing house design, even though they could have enough items to cram in a building for instance. In knowledge terms, it is even more dramatic. The reason? Information needs vary tremendously from one organization to the next. The culture, processes and data are very different, and software companies create a myriad of applications (houses) for one customer but packaging them as it could work for anyone. The result is a very long list of software options, and managers believing that one of those "solutions" will fix their problem as mentioned earlier. Now, if we lived in a world where no change happens, it might work for a while, but that is not the case. Your requirements today are guaranteed to change tomorrow.

With already made software (Especially with a fixed interface) , you will be working with another person's point of view, not your own. Ask anyone in IT if they would rather prefer to have people adjust to how the software works, instead of the opposite; adjusting software to the way things work in your company. You might hear, there is no way to do it, or you can't do that with software. This is because, once your house is built, making changes will affect many other elements, damaging the integrity of the original design. Change is what software packages are not designed to handle easily today because many of their inner core parts are fixed or unchangeable. . Customers are not allowed to change them as they see fit. That is the power of Open Source. Open source is software written in open and clear form, letting anyone modify it as they see fit. The condition is that those modifications must be made available back to the community.
If you need a bigger and faster plane, you do not take a 707, and bring bigger wings, engines, etc. You design a new airplane, because you want to maintain integrity and make sure it will hold your requirements. Integrity is lost once you start making major changes to a solution design.

- L-V-S-P-R-S: The connection between Leadership and Solutions.

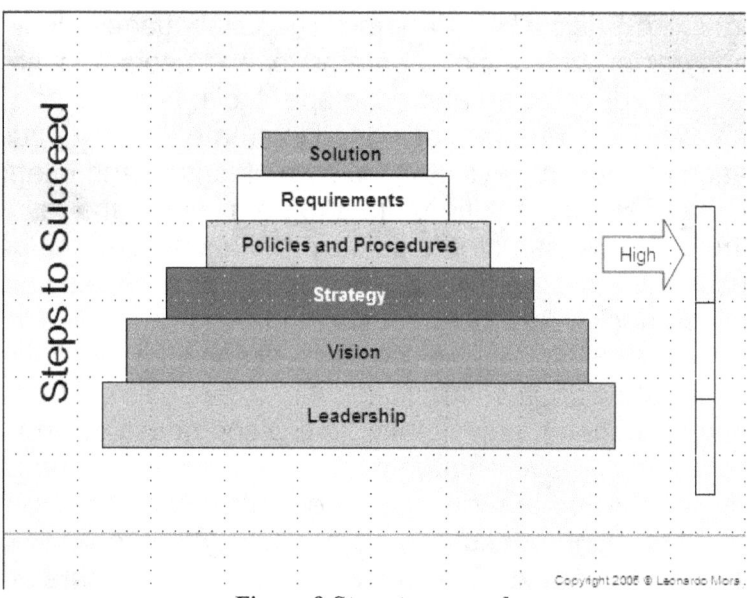

Figure 9 Steps to succeed

Another aspect to the solution is in understanding the path from Leadership to solutions. Leadership->Vision->Strategy->Policies->Requirements->Solution:. Leader's primary role is to create a Vision. (Insert reference) That vision is the point where you want to go. A set of strategies are then based on the vision below about how to accomplish the goal. Then, you move to transforming those strategies into detailed policies and procedures (or processes) about how users should be guided (In knowledge based systems), what are the rules in managing information, and what are the exceptions to those rules , responsibilities and accountability for results, and any special situation you can think of. This set of instructions should be revised and updated often with feedback from users to reflect reality once you put it in practice. The result can be then translated into a set of requirements that will guide the design of the solution in a NON TECHNICAL LANGUAGE. Then and only then, you can call the technical experts to help you find out HOW to build the "designed solution".

Successful thinking model: By successful --as we said before-- it means learning to define the why, what, where, how in a specific order ---a.k.a defining a plan based on the vision. It means getting rid of the "Quick fix" mentality and into clear definitions of what and how the solution should look like and work.

In figure 4, notice that if the solution you are trying to create is not based on the vision leadership created, then you are most likely creating something which will have no metrics for success: "***The fundamental reason why projects fail***"

Chapter 3

The Learning Organization

As mentioned before, to successfully accomplish one hundred percent collaboration through technology, it can only be possible if the Learning Organization's theory explained by Peter Senge's book "The Fifth Discipline" is put in practice. I make a short overview of the main disciplines and what a Learning Organization is about as explained in his book. I will leave to the reader exploring more in depth the book and any other material, which I find crucial to being successful in bringing collaboration to the working place.

Learning Organization Definition : "Organizations where people continually expand their capacity to create the results they truly desire, where new and expansive patterns of thinking are nurtured, where collective aspiration is set free, and where people are continually learning to see the whole together." Peter Senge.

Some companies make an approximation to this definition ,GE for example: ***"The second management concept that has guided us for the better part of two decades is a belief that an organization's ability to learn, to transfer that learning across***

its components, and to act on it quickly is its ultimate, sustainable competitive advantage. That belief drove us to create a boundary less company by de-layering and destroying organizational silos. Selflessly sharing good ideas while endlessly searching for better ideas became a natural act." Jack Welch
Former Chairman and CEO, General Electric Company

In summary:
The organizations that will truly excel will be the ones that discover how to tap people's commitment and capacity to learn at *all* levels in an organization. Learning organizations are fundamentally different from traditional authoritarian "controlling organizations."

To achieve it, you must learn the five disciplines needed to transform the way we think and work today.

Personal Mastery: Is the discipline of continually clarifying and deepening our personal vision and focusing our energies to develop patience and seeing reality objectively. People become committed with their own lifelong learning. **Mental Models:** are deeply ingrained assumptions, generalizations, or even images that influence how we understand the world and how we take action. Includes the ability to carry "learningful" conversations that balance inquiry and advocacy, where people expose there own thinking effectively and make that thinking open to the influence of others. In plain words, is to have an open mind. **Building Shared Vision**: The leadership skill most understood for thousands of years that inspired organizations has been the capacity to hold a shared picture of the future we seek to create. When there is a genuine vision, people excel and learn, not because they are told too, but because they want to. **Team Learning**: How can a team of committed managers individual IQ's above 120 have a collective IQ of 63?The discipline of TL confronts the paradox. It starts with "dialogue", the capacity of members of a team to suspend assumptions and enter in a genuine "Thinking together". **Systems Thinking**: It is the discipline that integrates all , fusing them into a coherent body or theory and practice. Without it, there is no motivation to look at how the disciplines interrelate.

The 7 learning disabilities

These points are for you to reflect the reality in today's corporations and how limited we find the model and framework under most people have to work with. It all starts with one word: *Responsibility*.

Seven learning disabilities

1 - I am my position

When people in organizations focus only on their position, they have little sense of responsibility for the results produced when all positions interact.

2 - The enemy is out there

It is a by-product of "I am my position". When actions come back to hurt us, we misperceive it as externally caused.

3 - The illusion of taking charge

If we become more "proactive" and aggressive in fighting the "enemy out there", we are reacting. True proactive ness comes from seeing how we contribute to our own problems.

4 - The fixation on events

The primary threats to our survival, both organizations and of societies, come not from sudden events but from slow, gradual processes.

5 - The parable of the boiled frog

Case for "Fixation on events"

6 - The delusion of learning from experience

Primary consequences of our actions are in the distant future, which it becomes impossible to learn from direct experience. Dilemma in organizations: We learn best from experience but we never directly experience the consequences of many of our most important decisions

7 - The myth of the management team

Maintaining the appearance of a cohesive team. They seek to squelch disagreement. Most managers find collective inquiry inherently threatening. School train us never to admit that we do not know the answer, and most corporations reinforce that lesson by rewarding people who excel in advocating their views, not inquiring into complex issues. (When was the list time someone was rewarded in your organization for raising difficult questions about the company's current policies rather than solving urgent problems?). If we feel uncertain or ignorant, we learn to protect ourselves from the pain of appearing uncertain or ignorant. That very process blocks out any new understanding which might threaten us. The consequence is what Argyris calls "skilled incompetence"– teams full of people who are incredibly proficient at keeping themselves from learning.

Chapter 4

Technological platform : A possible solution

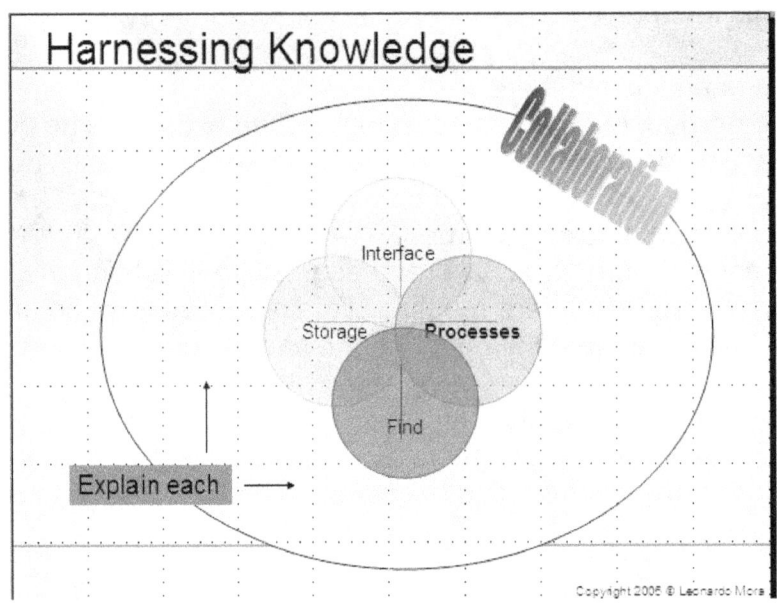

Figure 10 The vision

In Figure 5, I took the knowledge concept of data and processes which are the key elements we can translate into technology, and embedded them in the model as storage and processes. I call it storage, because all data is stored somewhere, but the way it is stored varies depending on the type. Processes handle not only processes defined by the user like purchase order, supplies request, etc, but manages the system processes as well. It means that the process engine should be able to manipulate any kind of process or operation.

The Interface

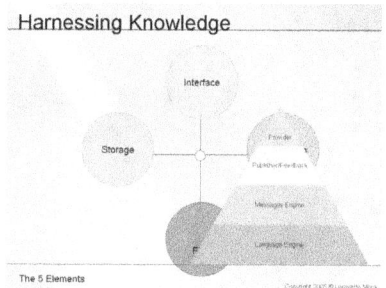

What should be our must important document? Which document should appear at the top?

How do we translate all the theory and definitions into something we can work on? Here is the solution.

The Interface engine is composed of (from bottom up) the language engine that controls the different languages the application handles. The messaging engine controls error, warning and information messaging in the interface local, as well as remotely on the client.

The next level is the Publisher/Feedback which has the ability to provide input and feedback through surveys.

Last, we have the Provider engine, which allow us to work on multiple interfaces .

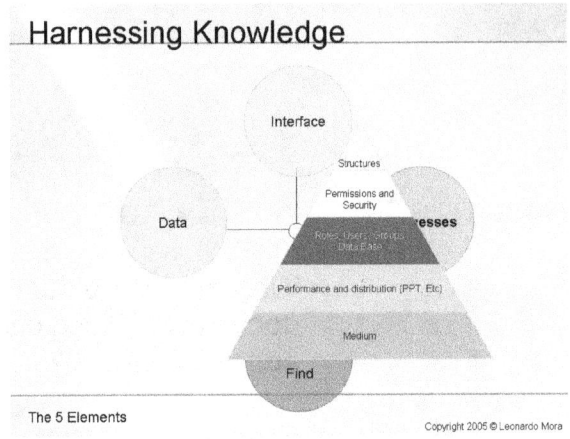

Figure 11 Data pyramid

Data

Data and it's storage begins with the medium. The engine controls the different devices that can hold data. The next layer controls performance and distribution, basically deciding where to store the data and how distributed it should be.

The roles, users, and groups, it is a model that tends to be very flexible and not rigid. It is a very flexible yet intuitive way to handle permissions in the system. Then we have Permissions and security and last are the structures which you build to provide navigational access to the people. That structure basically can be a physical repository plus a functional taxonomy and a logical listing.

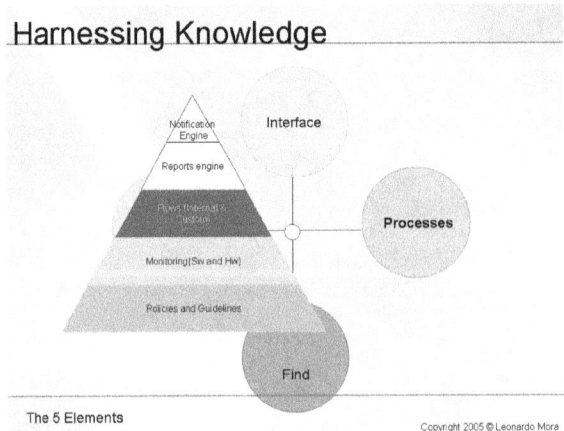

Figure 12 Processes Pyramid

Processes

 In this diagram we move to the third element: Processes. At the bottom of the pyramid, we find something called policies and guidelines. Effectively we want to convert every policy in your company to be part of the database. In order for them to work, we need a monitoring system which monitors software as well as hardware; its own, and others.

The next layer is called Flows (Internal & Custom). In here we can configure traditional and very sophisticated workflows. **The reports engine can be used to format all monitoring and workflow processes and generate professional reports.** At the top we have a notifications engine. It serves to send emails and instant messages to people who subscribe to a specific event on the system, either with processes or data elements.

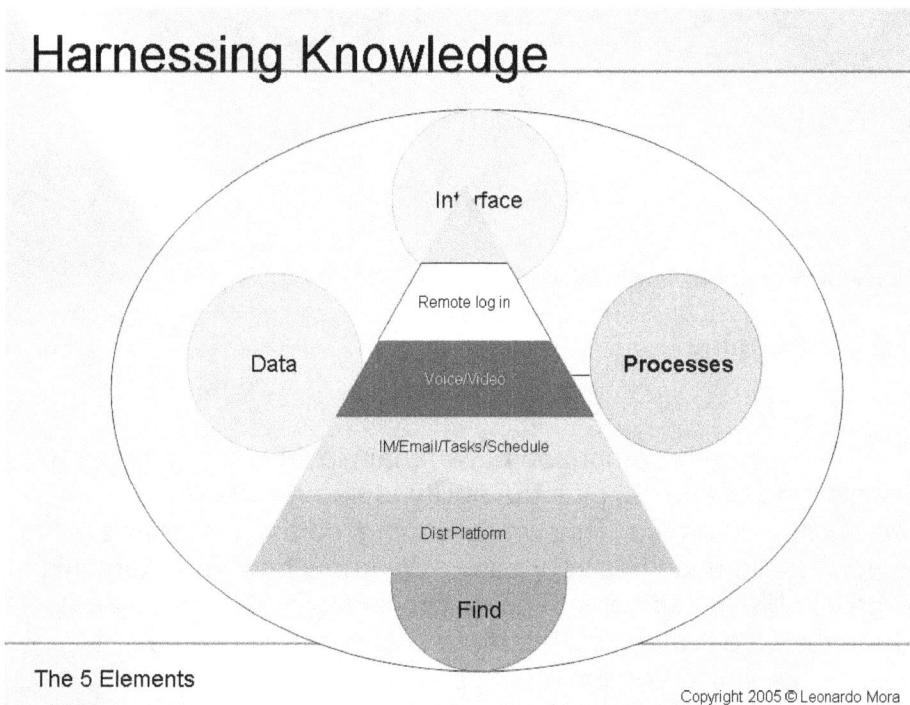

Harnessing Knowledge

The 5 Elements

Copyright 2005 © Leonardo Mora

Figure 13 Collaboration Pyramid

Collaboration

In the collaboration section we start with a Distributed Platform. The idea is that this is the client people will use, and in order to be flexible, it needs to work both central and distributed.

IM/Email is essential and integral part of this system. It merges with the Notification layer under Processes. The Voice and Video layer is an optional item, to provide multimedia capability to the client.

A task scheduler is built in for tracking various tasks and it could be used with email

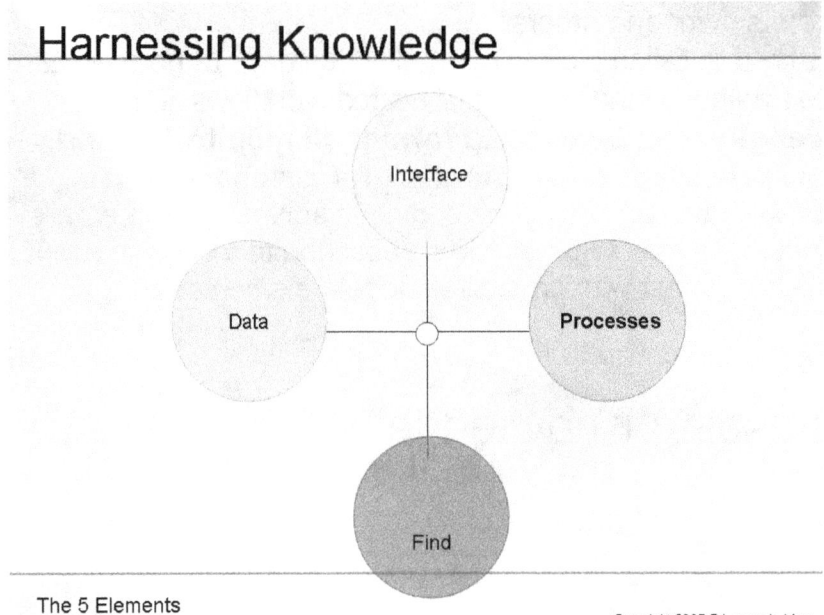

The 5 Elements

Figure 14 Search combined with spreadsheets

The Find Engine

The Find engine is comprised of two main layers, the first with a Google type of engine, plus the ability to see results in a spreadsheet interface. This way grouping, sorting and rearranging search results becomes very easy. We know how important this is, being able to find something quickly is essential to repositories, where the bigger they get, the more costly is to index them. I imagine a solution combining the power of a Google type of index, with spreadsheets, as the ultimate solution.

Chapter 5

Migrating from A to B

We've reviewed how to solve the knowledge problem – Collaboration, teams, etc--. But we haven't talked about what it takes to move from here (your current situation) to there. It is another 800 pound gorilla problem in large organizations. With the number of terabytes increasing exponentially each year (And petabyte hardrives already in the market), it will be no easy feat to move a giant data monster from it's current place to your envisioned solution in the future if we do not come up with a big plan and an automated system to do parts of it. No solution can solve everybody's problem, but the idea is to create a platform flexible enough to accommodate the different design solutions organizations come up with to generate and store rational knowledge.

Let's suppose you understood everything I said, and now you ask -- what's next? If you thought the tough part is over, think again. This section talks about the issues you will face when you look at your current situation to be migrated into the envisioned solution of tomorrow.

What are the issues?

Silos: Information is created based on the company structure. If it is a Silo, information will be in silos too. What is a Silo? It means that your company is compartmentalized, that your departments do not talk to each other, that everyone wants to keep their information from others.

Maps: The consequence of silos is that you cannot create an information map. This means links between files are non existent. We want to be able to "picture" the way information is connected to one another. Without a map, we can't.

Importance: Another consequence of silos is that there is no hierarchy that tells you the importance of documents. Many times you need to make a call to find out how important it is.

Permissions nightmare: In many corporations, there is no overall strategy for easy handling of permissions not to mention digital information. In many cases the non-official strategy is "Individual" permissions which means doom for any migration effort which translates into a very hard and time consuming effort to deal with access. A good way to deal with permissions is a mix between groups and roles as they are more related to other functions through out the system. The issue then becomes the synchronization of those groups and roles in your system B or target solution. Remember this when you plan for a migration. Massive amounts of information: We have today a lot MORE information than yesterday (Big companies are already in the terabyte-petabyte range). So the problem is not getting any smaller. Been able to massively copy, move, tag, log is key to any effort of consolidating information into a flexible platform. This section (Migration) has to be able to deal with errors, inconsistencies and unknowns in a way so that it groups them for later review. As when you look at your personal information, this is always a good time to check for information you want to delete. So filtering and sorting is crucial to let the owner find and select documents for deletion. That deletion can be temporal meaning it does not really delete until certain later date. Because the web was created on the basis of linking documents one to another it is easier to find information because is linked. In a corporate environment we do not have that. It comes from Silos as already mentioned. We need to mimic the web in our internal structure to build the web of internal information in a corporate environment. How? Making the linkage of documents the center of all effort. **Information unconnected ness**: Many corporations ask them selves why is easier to search in the Internet something that in their corporate network. After thinking about it, I found that the simple answer lies in something called *hyperlink*. Information in the web is linked one another. Corporate information is not. Search engines like Google, can look at the links between pages, and make a relevant list of web pages, because hyperlink take into account human perception of what is a good source of information and not, through the links they include in their HTML pages. This environment or mechanism is non existent in the corporate world. Excel, and word and other documents are not consistently linked to one another, so finding documents the same way we do it on the Web, is clearly not the successful as long as the documents remain un-linked.

When faced with a migration, it is clear that you want to have a clear picture of your solution. In one project, we did not have a design ready by the time we started testing a tool to move documents, and it proved very difficult, because many decisions are based upon the way it WILL look like, and the process users will follow to manage the information.

In the last 10 years projects involving knowledge tried to implement common methods that did not include the importance of people ;what I have found is Knowledge Projects are very sensitive to people (The ones producing it), and the approach you

.

take must be very different from any other type of projects. The new approach must involve the people it will affect and how you motivate them to use it is key. You need to consider "The Human Factor", in order to be successful. Also, there is a lack of understanding of where you are (which will scope the effort to move away) and where you (the solution design) want to go, and because it is so complex in information terms to do it and it involves all aspects of the business, it is a very challenging problem.

If the bottom line issue for you is bringing people in to share in their knowledge, then the following recommendations are for you.

The way forward

Main strategies needed to create and implement a solution are:
1. Know were you are: what is your current situation? .
2. Where do you want to go? Knowledge collaboration? This is your Key Goal . To give users multiple access points to information giving them open access to all

Figure 15 Horizon

resources and meeting places as IDEO does in their premises to foster knowledge sharing. Having just one way of dealing with documents and other objects is not recognizing that

we all people work and think different, (the reason why is important to involve users). In other words, different ways of motivating employees into sharing knowledge is better than trying to do it with a single tool or technology.
3. Less is more. In KM applies like a big hammer. Web 2.0 apps that were born on the web and not in a business are proof of this. What ever you create, keep it simple, and give access to as many people as possible.
4. Creating policies and guidelines to host a Document Repository. Many companies do not have a policies and guidelines document. Make sure you create it, and if you make it available on the web, people will use it, and keep maintaining it.

5. Fitting to the Organization. The solution you design so that the tools are conformed to the organization's culture and processes will succeed, not the other way around. This means all too often companies look at software tools not aware that the tools are the ones that need to conform to their way of working, not the opposite. Yes, we may have to enhance our processes, but not all at the same time; it could be very distracting and time consuming to try tacking data and processes at the same time. You must first tackle data, and then processes, because processes sit on top of data. Processes need organized data, so that it can be defined. First we organize data and then we create a process that uses the data or repository.

6. The system should match the organization's structure. By modifying the way you are structured, that on itself is a major change. You should be very careful on how to approach changes, because remember that you work are part of a system, and changes in the structure are critical. The suggestion is to try and work with what you have, and once you implemented the technology, make small changes towards a better situation.

7. Business processes must be matched as workflows[19] especially critical ones that affect productivity and motivation.

8. Fostering continuous practical training. People need to be reminded once in a while about certain features (and policies) of a given organization.

Strive to keep employees up-to-date at all times. Training people to "learn" how to work with and use a knowledge system does not mean the system "will" be used. Studies show that people will forget mostly everything by the 6th day their training has passed. It is important to coach them and hold their hand in the beginning so that change does not scare them enough to prevent a successful transition. Creating an information map: Knowing how many, how big, which types of files your organization currently holds is crucial for you to measure implementation efforts on a Repository. Knowing were point A (Where you are) will better guide you on the detail required at this particular project. When you create a map, there are policies required that get created along, like how many

---

[19] "Scientific workflows found wide acceptance in the fields of bioinformatics and cheminformatics in the early 2000s, where they successfully met the need for multiple interconnected tools, handling of multiple data formats and large data quantities. Also, the paradigm of scientific workflows was close to the well-established tradition of Perl scripting in life-science research organizations, so this adoption represented a natural step forward towards a more structured infrastructure setup. Business workflows are more generic, being able to represent any structuring of tasks, and are equally applicable to task scheduling within a software application server and organizing a paper or electronic document trail within an organization. Their origins date back to the 1970s, when they were purely paper-based, and the principles from that period made the transition to modern IT infrastructure systems." Wikipedia, June 5th 2006

files do we want to keep per type, or how should we delete obsolete files, etc. Your process for transitioning point A to point B is as important as defining the target. Create Rules and guidelines for grant related information and other e-documents for users to follow. It is essential to provide users guidance for the sake of system efficiency. Without the BUY IN from users, you will not be successful in the long run. Guaranteed. Your users are the most important part of the entire process. You need to involve as many of the most visible as you can. Without them your efforts and cost will increase exponentially. Most companies fail to capture their knowledge because there is no Vision and Strategy about rational knowledge, no information map or people assume that there is one but in reality there is a paper document policy not fully translated into the digital world or no automated policy system that helps you monitor collaboration. Thinking that in today's massive libraries you can manually verify if policies are working as planned is not realistic. Files and process organization requires careful planning and thought. Streamline document creation and storage making it simple yet intuitive to work with. Increase accessibility through search tools. These lessons about how things should develop in a knowledge project come as a conclusion from past projects that did not have a clear method for repositories, interface, search, and how to handle workflows and processes.

Repositories – Map

The following sections, will address specifically how you implement the concept of structures in a repository. If you are not part of a similar project, you might want to skip to creativity and innovation section.

The first thing you need to do when trying to go from one point to another is to define where the initial point is, or A taking A-B. This means creating an information map to know where everything is, and how much effort and materials will consume the task of migrating.

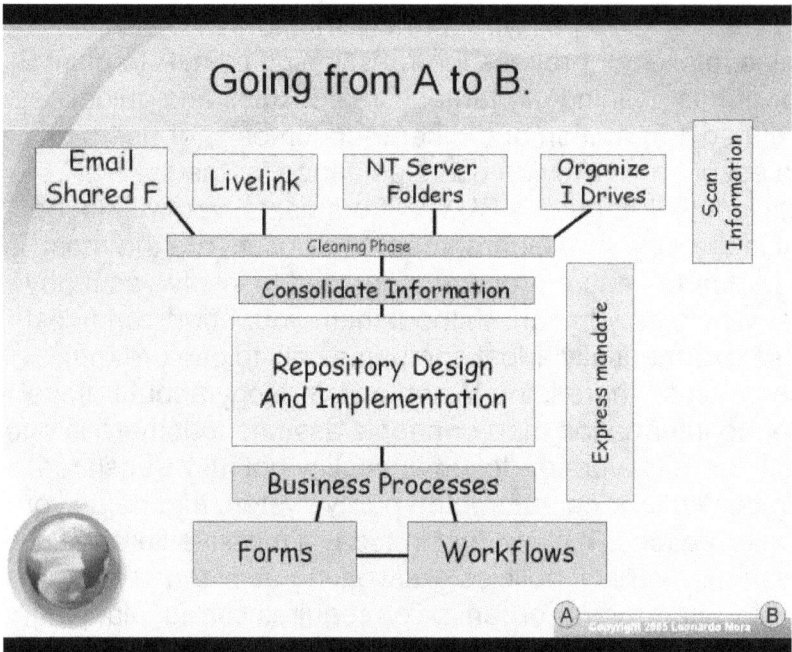

Figure 16 Going from A to B

The green section is about identifying the different systems holding your information. Identify possible conflicts with duplicated documents and/or special needs in processes requiring more than one group that are to be migrated at the same time. It can provide valuable statistical information about the growth trends, habits, and current activity in all your systems.

In this diagram we follow the elements in our framework Data-Processes and eventually Collaboration. It is always recommended to have a cleaning and consolidation phase before designing your repository; it can give you a fresh start instead of trying to do it later down the road.

After many implementations is safe to say you will need a strong mandate for this sections to be carried on by your employees. So do not be afraid of asking for this mandate to your CEO or president because this is where leading the way makes a very important fact for the project to succeed.

What I am recommending in this diagram is to have a repository first and then your business processes. Why? Because many business processes may require you to have the repository at hand prior to implementation plus it makes it easier for you to do it while you are doing the repository design first.

In my experience I have found out that some groups (I wont name anyone) have the expectation that a consultant will come with a magic formula to convert the mess they have into something clean and pristine. I had the sense that for some reason they think "one solution fits all" will work in their company. Another type of thinking is asking consultants who else have the same solution or implementation. In reality, these projects are each very different to implement because each company or institution is different, how? in the way they conduct business, the way they store information, in their culture. What you need is to assess each organization and come up with a plan based on the assessment and not from some past experience that seemed to work which is important but it cannot be assumed will work for everyone.

In one of the companies I worked for, I could see clearly what happens when you try to implement a repository, but there is on clear mandate from the CEO: It dies overtime, at the beginning people might be excited, but slowly they will stop using it, first because the leadership is not demanding the use of it, and second, we had more than one choice of system (email, personal folders), so not only they were not told to do it, I had an option of deciding which one I would use.

What is a repository?

What is exactly a repository? From the dictionary we read "a facility where things can be deposited for storage or safekeeping". In technology, a repository is a central place where you can store a variety of objects. One requirement is that the policies and procedures you envision for your company in the digital age need to be clearly defined. Now to the question –Do we make it centralized or distributed? To make thinks simple to manage, you want to centralize, but if you have multiple locations (As a multinational company does), then you need to combine it with some sort of distribution, so that each location can access rapidly your data. So my take is that the solution must meet both schemes but not at the same levels. The other issue with central repositories is that not every one is online all the time (e.g. Airport-airplane,

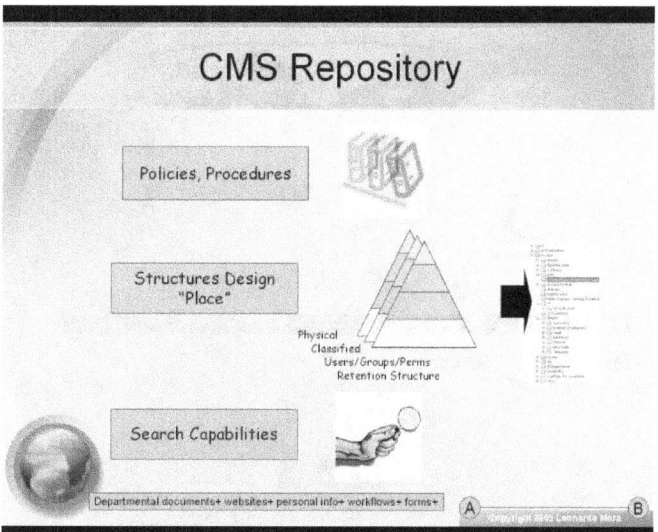

Figure 17 Repositories

train) or at least not today, maybe in the short future yes as we see these areas been converted into online hubs.

As it was said in the first section, policies and procedures are a must for you to develop correctly a repository so make sure you have them (they do not have to be complete and extensive) before going any further. I strongly recommend automating those policies.

The structures in a repository can be divided into 3 groups.

Physical Classification, Views, Permissions.
Permissions can be further subdivided in Users, Groups and Roles.
The way we find what we are looking for relies heavily on how good your search technology is combined with how flexible is the tool to handle the results in different ways. I call it the FIND engine. Combining search results with spreadsheet functions gives you a lot of power. Imagine having results from Google or other search engine in a spreadsheet format, where you can sort, expand, filter, etc. Those familiar with Excel and working with this layout, would understand the kind of power I am referring to. The problem with search is that not all "High quality" pages are always in the first batch of results. And depending what you are looking for, this method might help you greatly I finding what you are looking for.

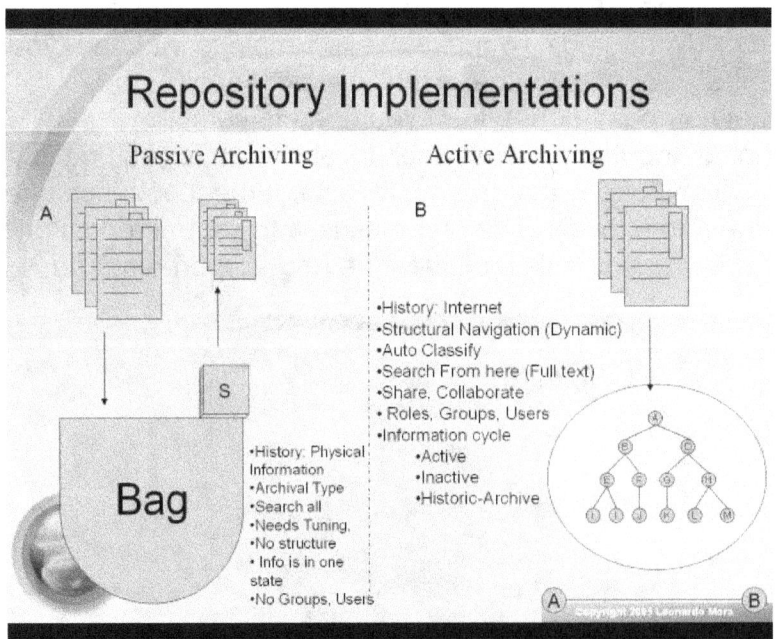

Figure 18 Two different mentalities

Repository types

Historically and after observing the way people see things, we have on one side the Librarian view which we call it Passive archiving, where you have books and publications (Physical items typically)stored in a single place (Public and private libraries follow this method), and you have a single database to find the title of the book, author, genre, etc. Its features are: Information is in archival mode usually for ever; when searching for a topic or author you are always searching the entire collection, information does not have a cycle, needs tuning when classifying records and it does not posses a structure. Internet search engines are a good example, when you search for the word knowledge, they display thousands, and even million hits. To go deeper into the results it's as hard as finding the needle in a haystack.

 The limitations are:

There is no way to find topics inside the material or full text indexing. Searches can be long and complicated. There is usually a "middle man" that is the expert in searching, called the librarian. There is no permissions structure.

On the other side we have Active Filing, which came about with the Internet. The features we find in it are: There is a structure (Navigational) and it can hold more than one like classifications, user permissions, etc. It is more dynamic, we can create cycles so that information goes from one state to another, Active-Passive-Archival. In this form we have a lot more interaction from people, sharing and collaborating on office documents. There is usually a full text index and you can search for information based on the structure. Classification is usually the key for handling document types and their retention. Ideally, with a retention scheme, the system can move back and forth information from one side to another. We would like to have this system function with the two sides , the passive archiving handling the bulk of data and heavy storage, and the active side handling the most used and active documents for quick retrieval. Its limitations are: 1- Not good for very big containers. 2- More complex to build.

In passive mode, we use something like a bag of candies; we throw every candy in, and expect to find them in one place. So it is central by nature. These posses a problem for multiple locations, where you need to replicate the data. (The active scheme works well for combining central with distributed approaches). Working with our bag, requires deep knowledge on the tool we use for searching inside the bag. Why? Because the results might not give us exactly what we are looking for. Results can be extremely long and complicated to filter. You might not have the ability to change the resulting records in one shot; this again depends on the tool and the vendor. Some vendors like working with you, some do not even pay attention, so you have to foresee how much support you can get or buy.

In summary, libraries are a good tool only when the records and fields are static, or you do not have multiple types of documents.

There is a need to reform the way libraries are seen today so that they can include a more flexible approach, the active mode I described above. This way, we can integrate both into one system, and the end goal of having this methodology is to be able to cycle the information.

This means that when information it's created, it is active for a period of time, then it falls into passive mode depending on the interest and activity and if obsolete after a given time –say 5 years) then it should be deleted if legally permitted; some records will have to be kept permanent then it should go into archival mode in a media that lasts many years, and depending on the legal requirements for retention, you can store it for a number of years or for ever. Media used in big jukeboxes (used to handle the bulk of data) can survive for many years, some to 100 years.

Basic structures

There are 4 basic structures that handle everything in a repository. 1. Physical 2. Classification 3. Logical 4. Users/Groups/Roles We have the folder structure

where we could follow the way the organization is structured to implement it.

Then we have a classification structure which provides us with a

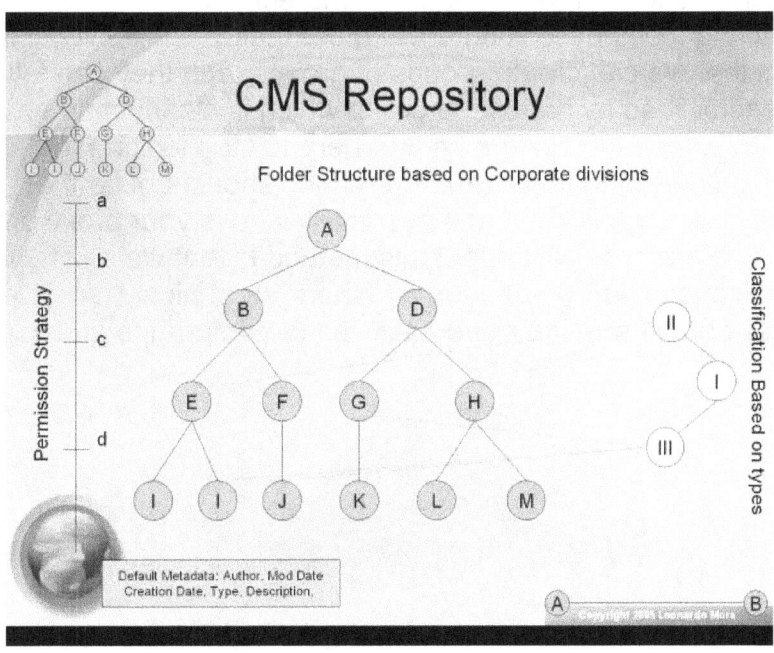

way to customize the various types of files we use. A quick

example: In legal documents you handle much of the bulk using MS word format. So only classifying by the type of office file would not give us any meaningful benefit as in this case mostly all are of the same type. We need to create types based on the documents themselves, like contracts, letters, approvals, etc; each type can define the retention period and a set of rules the owner requires like for example which medium you want it stored, or how long should it be kept in active mode before sending it to archival, etc. Custom fields can be added to a type so that sorting and searching can be more powerful and flexible at the same time. The folder view should allow you to view not only its contents, but any classification type defined in the objects it contains. This provides you with a mechanism of displaying each node differently depending on its contents. Music and math files should be displayed differently. The logical structure is targeted more to the "How do I do this" type of question. It does not reflect the groups in the organization but the most important tasks needed to be performed by teams of people. Examples: How do we buy something? What is the process of acquiring insurance? How do we close a deal legally? How do we ask for services? Who is the expert in IT and databases? All of these questions can be put together in a simple structure to help others find their way without getting into the intricacy and complexity businesses create. The way I've been learning how the last structure should work follows a simple rule. Things should be guided by the role more than the name of the person. So this is what it will look like: Groups Roles People Groups should have a setting where it tells the system not to go in to big loops by looking down more than 2 levels. What that means is that we can create groups of groups, and they can fall into a deadlock, so by creating a group which is clearly marked as a group only, then the system know where to stop looking for permissions (More on this later). Then roles should define the name, any rules applicable, and especially who is your proxy or backup. The major benefit about this method is that we can have the organization chart embedded into the system nicely, so processes can be created more easily based on groups and roles.

Classifications and structure

A classification tree is a structure where we divide information not from the unit or department where it comes from, but from the nature of the information, like contracts, purchase orders, letters, requests, financial reports, manuals, etc. Classifications are key because they can be used as the "controller" and to store additional metadata concerning each type, meaning many of the rules legal and otherwise can be consolidated in it's classification. Another aspect that can help you handle better documents is the flow of uploading them, or the number of steps needed to add a particular document. For Example, a Purchase order might follow a different path than the contract for legal services because it involves different sets of groups.

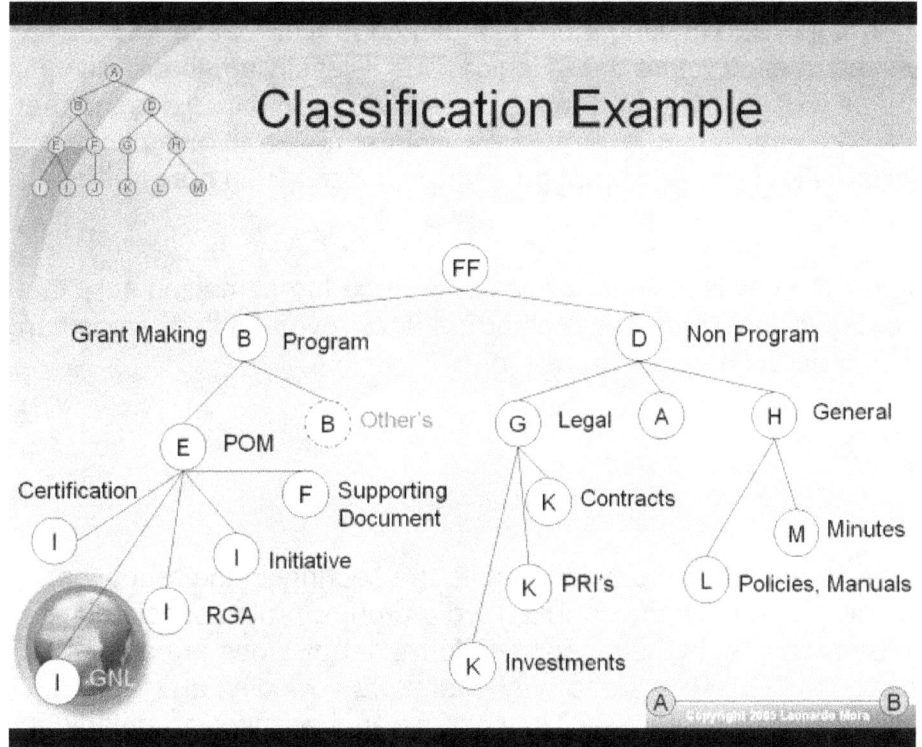

Figure 20 Example

We call retention to the rules defined for each type of information that you want to store. For example, a contract might have the following rules:
• Remain active for 1 year.
• Remain passive for 3 months
• Archive 10 years.
• Cannot be deleted without approval.
• Modifications should be notified to owner.
Information can be treated as a living thing, having 3 main stages called Active, Passive, and Archive or Deletion stage. With this method, information can remain up to date, be more useful, and the overall organization can improve the performance of the system.

I believe classifications can serve tremendously in organizing information and making more useful a central database, in parallel with a folder structure and a logical one. Classifying should give you as well a way to customize metadata (Additional fields) so that you can sort and retrieve more easily information. It gives you flexibility and power to configure your information in many different ways.

Again, the key is in your policies and procedures, making sure that a broad vision is set, and a clear mandate given so that teams can implement a complete repository.

Groups, roles and users

These 3 entities are used to handle the security in the repository. Simply put, each node should have a group or role associated to its permissions. In the most basic form, permissions are applied in the form of Guests = Read only, Members = Modify, and Coordinator= Full access. They can be applied more granularly but that should be left only to power users or administrators of the system and it (personal permissions) should be done in special cases or what your policy mandates. Permissions can be assigned in a hierarchy structure that allows more or less privileges. Every system I've seen did not take seriously roles until the modules for workflows gave or required users to define roles for the flows at hand, as many tasks in a given flow are more readable if we use roles than if we use a person's name or a group. Groups are important to organize people, and there should be a special type of group that holds other groups, but to prevent deadlock loops, we should limit it to 2-3 loops when the system is checking for permissions. Recursive groups can grow in complexity, that is why we need to put some safety measures to prevent the system from crashing or slowing down.

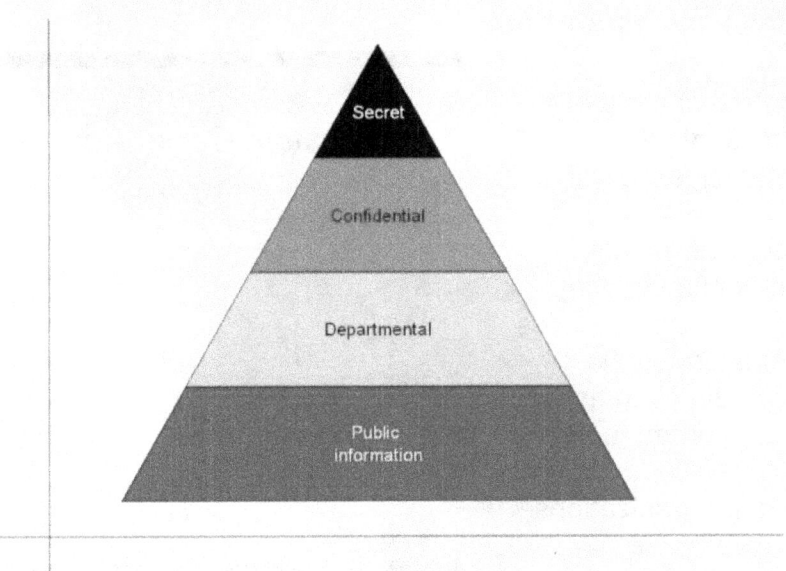

Figure 21 Permission Structure

It's important to note that there needs to be some kind of synching between these groups and roles with a central directory or at least at the messaging level. Ideally pairing it with a directory services automatically will save you from some headaches.

There should be three sets of roles seen as pools of people that can be assigned to each object:

Owner Or Administrator.

Member

Guest.

To each of those we should be able to add any group, role, or user in a temporal or permanent basis. Temporal should be set for 1 day, or 1 week. , and permanent should be allowed only to the administrator. A report on this assignments should be available for review .

The permissions should be set by roles. This means that the administrator has full permissions, the member can add and modify objects into the place, and guest has read only privileges. A special member can be created where permissions are set so that they can add documents, but they cannot modify any .

Creativity and innovation

It occurred to me that the way to handle the perpetual question of

how do we promote creativity in a system, would be to grant access to information in the opposite direction as we are accustomed to see.

In current organizations and specially the intelligence community,

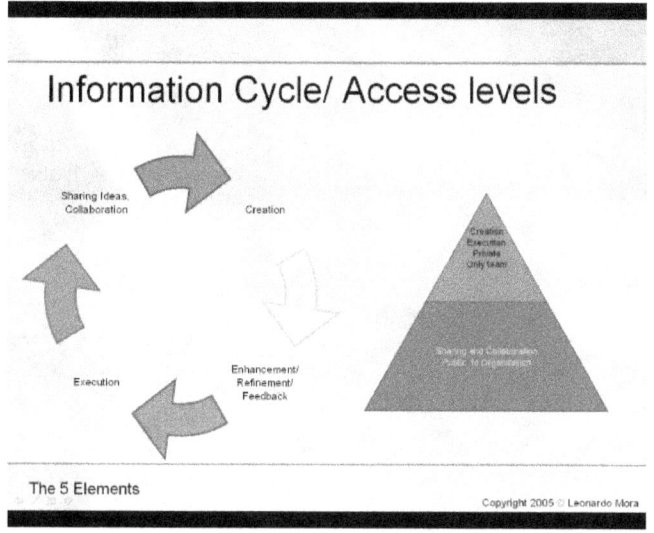

Information Cycle/ Access levels

Sharing Ideas, Collaboration

Creation

Execution

Enhancement/ Refinement/ Feedback

The 5 Elements

Copyright 2005 © Leonardo Mora

Figure 22 Cycle

the higher you are in the structure, the more access you have to critical, confidential, secret information. So very few people actually see what is really important. That might be a big problem. The higher you are from the ground, the less you can actually see , and observe any detail.

Yes, by granting more and more access to people who are :
1- Closer to the customer 2- Experts or heavy contributors to the system, we might generate the push needed for the "small" people to be more productive, creative , hence innovative. But based in the methodology reviewed earlier of "learning organization" the focus has to shift from individuals to teams. The entire team has to have the same access level to information. I am simplifying the structure down to two phases. The one where you create, modify and refine and then execute, we'll call it private phase. The moment you share the results of what you created, we'll call that public phase . A simple way for people to understand how you can create new knowledge in an enterprise.

Control might prevent innovation

What would happen if an organization decides to let go trying to control information? Or dare we say knowledge?

I imagine people would automatically feel more comfortable. I am not including in this sentence not monitoring what is happening. This means let things happen but make sure you make yourself aware on what is going on, so that you can actually learn and take action.

I was reading a book called "The little book of letting go" and it explains in the first chapter how we adults want to control everything, be right all the time and keep judging life , and that certainly happens to me. Looking at children, they are happy because they are not concerned about those three things. They are focused on today, now, and they have little time for unnecessary thoughts. In order to be truly creative we might need to let go control , and let things flow but monitor them, read feedback closely, learn and adjust.

We can create a friendlier system by "allowing" sharing information instead of blocking people , we can make monitoring and notifying the core functionality in the system so that a lot more can be done without having manual processes.

An example would be how banks perform a lot of their in house development. They do not have automated tools to check and test the software, generating a big amount of flaws and in consequence a lot of time spent into fixing those. They could resort to use automated tools to improve the efficiency and reducing cost because if we can write a law it could be " If you want to reduce costs, you need efficiency". Now days we need to automate more to reduce errors.

Collaboration in the future

I've been thinking what the future would bring in terms of devices allowing us to communicate and collaborate better. We need to create an all powerful media device that allows us to use any kind of current communications method. The main question I ask is how voice, video, email and chat will converge into something powerful yet simple enough to allow ubiquity and action anywhere we are. A cell telephone is the place to start. The way it is designed and specially the screen size and functionality are the most important factors. Digital screens that flex are a reality today. I can envision a device that has two sides and you can have a chat, conference, etc on one side, and in the other you can send, receive, share documents, text, etc and be connected to our system at the same time. I recently saw Nokia concept on foldable phones that can be even wrapped around your wrist. It would be extremely useful to have this flexibility. The future looks pretty exciting and positive.

Thoughts

All signals are pointing to a big boom coming in technology and the time to start companies might be very well now[20]. Article
I've always believed that the customer is king. Here is a quote I liked
 "A customer is the most important visitor on our premises. He is not dependent on us. We are dependent on him. He is not an interruption of our work. He is the purpose of it. He is not an outsider to our business. He is part of it. We are not doing him a favor by serving him. He is doing us a favor by giving us the opportunity to do so.
"

Open Source projects and software are key to what we are trying to achieve, as much of the things we need to do are simple, but simple does not mean is easy; we need to work from other's creations in order to innovate in a shorter timeframe. Maybe that is why many people cry foul when other's patents holders come claiming ownership of your project. Either way, it is very exciting and I look forward to the challenge of breaking new grounds and markets. "The combined effect of these trends is dramatic: For JotSpot it's cost was just $100,000 to get from idea to launch; for Excite, $3 million." from Business 2.0. That is a huge difference.

Organizations and new structures

In the future, the focus on hiring people should be not so much on their current skills and experience (Knowledge i+e) but their learning capacity and team work (A.k.a how well they collaborate with others).

To motivate these future workers, the way we reward them has to change significantly. We need to be able to measure knowledge contributions and impact in the organization by tracking them down. One way is to have the knowledge repository integrated with the payroll system. Then a rating system where team feedback can be collected could be used as a validation mechanism so that payments can be increased (or decreased) in a certain range based on people's contribution. Not sure how good can be, but it should make the salary go up as well as down so that people know the impact of their contributions , a strong incentive to be collaborative.

---

[20] http://www.business2.com/b2/web/articles/0,17863,1122935,00.html

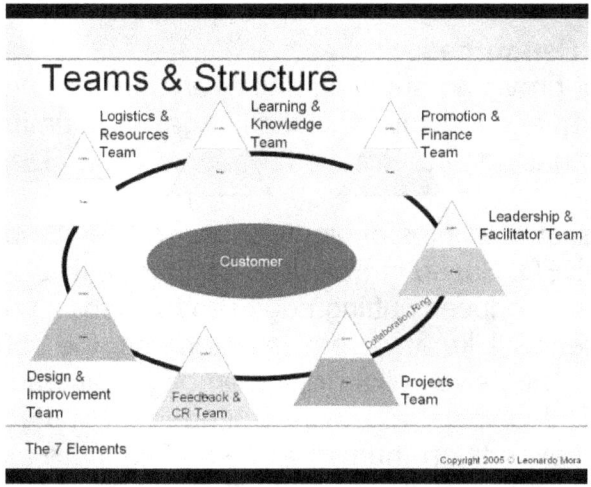

The structure of this organization looks very different from what you know. Here is a summary:

Teams Description

Business Promotion & Finance Team
Former sales, and financial team. Sales is no longer the focus, the focus is in the customer, and promoting a well thought and excellent product. Rotation: Projects team

Leadership & Channel Team
Overall leadership, inspirational and organizational Setting Organizational Vision Define main strategies Monitors performance and new projects Rotation: Feedback team.

Projects Team
Project teams developing products and services for outside customers and Inside Employees. Rotation: Promotion & Finance

Feedback & CR Team

Figure 23 New Paradigm

Creates surveys, forms, track, analysis of customer feedback. Works very close with customers to learn their needs. Conserve statistics and historical information for future reference. Applies same methodologies inside the organization
Rotation: Design and improvement team

Design & Improvement Team
Policy design, rules and processes creation. Reviews current surveys and recommends new and improved methods and processes. Rotation: Any and All Teams.

Learning & Knowledge Team
In charge of creating and monitoring all training and learning

aspects of the business. In charge of knowledge repository, sharing and collaboration of information. Responsible for implementing 4 elements of the Fifth Discipline. Rotation: Any and All teams.

Logistics and Resources

Responsible for providing anything and everything the business needs to perform and execute; Includes Legal, IT, Human resources, Purchasing, etc. Rotation: Projects Team, DI.

One of the closest examples of highly collaborative companies I have found is IDEO. Based in Palo Alto California, IDEO is a firm that designs and produces cutting edge products for a wide variety of corporate clients. "Like any creative enterprise, the $50 million firm depends on intensive collaboration among a diverse workforce of 350 industrial designers, electrical engineers, manufacturing specialists, and experts on "human factors." The more these people can cross-pollinate their talents, the better IDEO's projects will be. Yet, moving from concept to mock-up to finished project as quickly as possible -- getting together every member of every discipline and every project solely through conventional means is impossible. Instead, IDEO banks on randomness, using its carefully stage-managed physical environment -- to increase the likelihood that individual interactions will happen on their own."
From: Fast Company.com Issue 28 | October 1999 | Page 150 **By:** Paul Roberts

Knowledge in Organizations..

After a couple of wonderful weeks back home, I was talking to my good friend Orlando about this book and what I am planning (Create this powerful knowledge product), and struck me like a thunder when I said something about knowledge and organizations... I kept asking him and several other people(days later) this question:

Where does knowledge reside in an organization?

The obvious answer I've got was "In people's minds", so I had to change the question to something more specific "Where does knowledge reside (in a consolidated form) in an organization?". What amazed me is that nothing comes to mind like "your knowledge system", or DMS, or Intranet, no. The answer was in people's head. Note: Recent research shows an average of 80% institutional knowledge to reside on employees head. Bill Gates is targeting knowledge sharing as a big next step.(Zdnet , May 16th 2006)

To me the answer is simple but striking. Knowledge (As I put forward in this book) is in ... The policies and procedures (P&P) manual or book!. Or it should   be there. This comes back a couple of weeks ago when I was holding a couple of books about how to write good P&P's. Intuitively I knew there was something important to me to look at it, but did not know why. Now I understand and plan to make use of it more extensively. But the plan is not to create a manual. It must be a system that actively helps us to manage policies and procedures.

In a good , serious, ISO XXXXX type of Corporation/organization , decisions , processes agreed, etc, are or should be included in their Policies and Procedures manual (for us should be a system), because thick books or manuals to tell the truth are useless (have not read the first in many years), and there is nothing more boring than reading P&P's because it is an art to do it well (as writing a book) after reading manuals recently about how to do it right , and you need a proper structure to maintain it alive, as any other information piece. This is something we see many organizations ignoring or partially doing (Those with critical areas as information security). But still, it is no easy feat to read P&P material.
Now, what the problem is/has been, is how to effectively write, publish, promote and track those rules and processes.
My conclusion has been that it should not be a book or document, but rather an active system that promotes itself, and monitors other systems(Data and processes systems) to verify how successful those decisions are. Then it enables you to get timely feedback, so you can take action and adjust faster to a given condition.
"There should be truth in thought, truth in speech, and truth in action. To the man who has realised this truth in perfection, nothing else remains to be known because all knowledge is necessarily included in it." Gandhi.

project was me which today sounds really crazy to a PM planning a similar project (A minimum of 5-10 people is a good range). We struggled to figure out what was the path to follow, so a lot was trial and error, but let me tell you that this method is not for the faint of heart. It is by far the hardest way of learning and the most expensive. That project was probably the first in the world (1997) in size and scope dealing directly and specifically with KM with absolutely no expertise around (I think they scoured the world around for consultants) to guide us without any luck. They found none.

But for some reason the top guy up there believed that I was responsible of the problems the project was having. That did not make me feel too good, so I decided to start looking for something else rather than waiting for the hammer to fall on my head. I was sure warned about it. It looked typical management stupidity where you start looking for the scapegoat (Shoot the messenger syndrome) instead of looking at how the project is structured. Luckily for me, my boss and managers from the consulting company where absolutely clear that instead of guilty I was brave enough for dealing with a crazy project like that. In the end it was straightened out by a very competent project manager—the third in a row. My first big lesson in project management: Make sure you know what you are doing, if not, bring the experts or don't try it unless you are ready to fail.

Implementing the technology was exciting, new ways of dealing with information, figuring out how to best organize, capture, retrieve data, create workflow processes, etc. The number of things involved was growing and growing big in complexity. We were learning about the importance of interfaces, the users aspect of it, and how delicate was to migrate information from one side to another. There were unseen hidden links between groups of people that proved difficult to deal with without prior research. But again, I learned from the technical perspective as well as the business side.

After that, I've got an offer to come to the United States from a big tobacco company. I was very excited, and as always happen, the excitement transformed into frustration, the 2000 year US election proved to be a bump in the road for the papers that were delayed so much that the project I was called to join, dried out. After some calls, I landed a job as project manager for a big financial conglomerate. I had to move to a hotel and start working the next Monday morning in a different country, different culture, and make sure I produced results by the end of the week in a project with "again" unbelievable goals. Not recommended for the faint of heart . It stroked me the way their goals were set. They wanted to go "paperless" in less than four-five months. That is something bordering with craziness (first making the attempt) but I needed the job, and did not think too much about it. I have to say it was again very rough, as the project was already underway, I was coming in the middle of it replacing another PM, and they expected results probably in an hour, thinking that there was a proven method (Shall I say magic formula) of how to implement paperless projects in a snap. Today we know you can't possibly go paperless unless is a brand new company with a different leadership and mentality. For older companies you will always have a mix of paper and digital information.

Ok, so the reason the previous manager was removed, was because he did not agree on the way they wanted to move their information to this new technology. And he was probably right. But I had to do it anyway, so we went ahead and moved all files and folders into the system in less than a month. We did it, but with a lot of errors, because the tools used were not designed for that kind of massive uploading and there was no planning what so ever. Next, I found my self with another challenge; spitting out a fully detailed project plan with no lead time. I had no idea about how to come up with it because you cannot possibly foresee a plan without knowing where you are, their current state of matters, where you are going, and a team of people. In other words, you have to be there since day one as project manager to be able to figure out why, what, how, when, etc to do it and plan it right, but I did it anyway. I can think back now and tell you how crazy everything was. I do not know if heads rolled from the client stand point, but being a consultant kept you in a somewhat safe spot, at least in my case. Luckily they knew that just bringing in a new project manager and doing the same thing would not cut it; another hard lesson in the pocket but very valuable. I was sure learning the hard way how NOT to do Knowledge Management from customers and companies shooting up in the air. That was probably the time I started asking my self why they were doing it like that, and began to step back. I guess that is the way we learn about new things. We fail and fail until we learn a new way that works better.

After a couple of other small jobs 9/11 came and we got a (Sadly) 3 month vacation as everything and I mean every single project was shutdown. When I've got the offer for interview in New York later that year, everybody in my family had the terror look saying to me "Are you sure you want to go there?", I thought to myself that probably the best place to be after 9/11 was precisely in NY, and there could not be any other safer place on earth.

In NY I was offered a two year contract, so we decided to move from our temporal residence in Maryland. The challenge: learning the inner working of a non-profit organization and fixing-guiding them through the tough KM waters. My first impression was that the pace and ways of working coming from the "outside" world were quite different to how the non-profit world behaves. Everything moved a lot more slowly. It was hard at the beginning but I adjusted and figured out the way you could move around and get things done. The technology was fixed, and the users where retrained with a different approach. It proved successful in the number of users using the technology, but at the expense of time. Still, many of the factors affecting KM applied here. It took a long time, as I was the only resource doing everything. I soon learned the value and power of having policies and procedures in place when there are none. Meanwhile I delivered a seminar on KM and in retrospective I realize how messy it must have sounded, but still today, I get people to my seminars that want to know what KM is all about.

It has been 10 years now, so let's look at the lessons from the past and how I looked back to try finding a clear picture of what is wrong with KM and the definitions behind. Here are the lessons I've learned so far.

Appendix 2

Truth , Trust and Belief

Epistemologist's theory on knowledge says that knowledge is based on beliefs and truth. T Williams proposes the opposite, belief is based on knowledge. Clarity in concepts is based on language definitions (A.k.a The dictionary). The Aymara tribal group in the Andes mountains in Chile, Peru and Bolivia, believe that the future is in the past, and that the past is in the future; A reverse concept of time. Their "truth" is based on their concept of The Aymara language "nayra," basic word for "eye," "front" or "sight," to mean "past" and "qhipa," the basic word for "back" or "behind," to mean "future." So, for example, the expression "nayra mara" – which translates in meaning to "last year" – can be literally glossed as "front year.". In all other cultures it is based on locomotion and orientation of bodies, that sitting in front ahead of oneself, and past in the back or path . It means, that given their knowledge (information + experience) they can make such conclusions. Same as when men believed the planet was flat. It was true until someone proved the theory wrong (Which led to America's discovery)

What about truth? In my experience, you can't talk about truth without including trust first. Trust is the relationship between Sender and Recipient (Communications theory). Trust defines how valid something is to be received. How do we know if Peter is saying the truth? That depends on Peter words and actions in the past. If he in the past has proven to be not trustworthy in his words and actions, then what Peter says today will have a less impact and credibility on my mind, because his track record has not been consistently trustworthy. I might consider true what Peter has to say if I trust him. This includes not only his words but his actions. We can talk one way, and act in another way. It sounds too familiar. What is the impact of trust: According to Watson Wyatt Worldwide of Bethesda, MD "...companies where employees trusted top executives posted shareholder returns 42 percentage points higher than companies where distrust was the rule." **Walk your talk**. Trust will only come alive if leaders reinforce their words with actions. They must be the role model of everything that they want to have happen in their organization." Wolf Rinke. He adds "There simply is no shortcut to developing trust with another human being. It can't be done via the Internet, voice mail, faxes, or other electronic media. It requires personal contact."

Appendix 3

Learning definition; It is the process of acquiring knowledge (Information for our purpose) or skills through study, experience or teaching.

How do we Learn? I will list them in the order we learn since baby stage.

- Through Imitation
- Through Experimentation/Practice/Experience
- Through Teaching
- Through Observation, Environment, situation.
- Through Study (Reading, memorizing)
- By Touching, Sensing (Experience)
- By Listening (Experience)
- By Smelling (Experience)
- Through repetition (Experience, Memorizing).
- By reflexioning. (Mind only)

Every process is a learning experience, and every experience is a learning process.

Appendix 4

The solution to a vendor problem

The way you can shield your company from what happened in the case reviewed in page 13th, is to build a layer on top of the vendor software (what I call normally your actual solution), so that at any given time you can make changes without affecting the look and feel, plus the methods and processes you built in

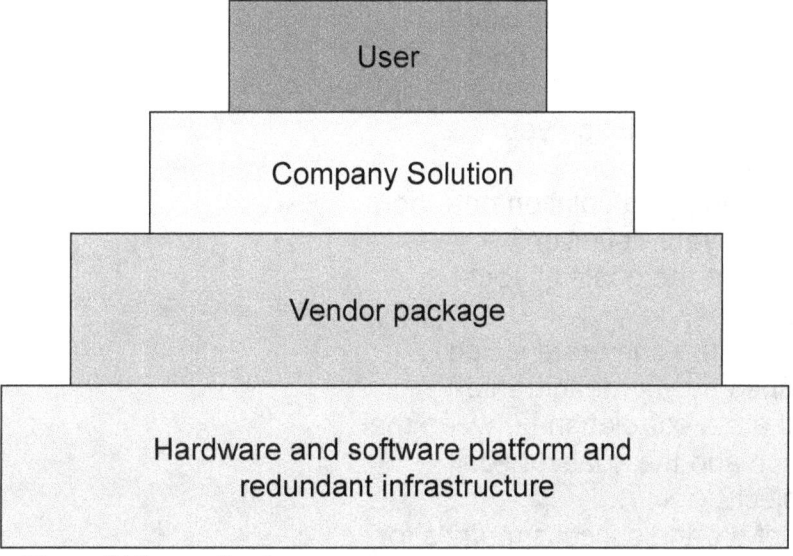

Figure 24 Separating layers

the application) or in the opposite case, shielding yourself against changes initiated by the vendor, who may affect you heavily. Figure 15 gives us an idea. When I mention redundant infrastructure, it includes power supply. I have seen cases where the entire company came to a complete stop, because when power failed, the back up system did not cover all users PCs. Without redundancy, you have a bigger problem than just implementing knowledge tools. The vendor package sits on top of your infrastructure, and the company's solution with all your policies and procedures, look and feel, etc sits independent from vendor software. If the vendor makes changes , you are not affected by them and you have ample choice to make a decision when upgrading software, when to do it, and if you want to do it at all. I should say end-user sits on top. As you see, the end user is king, so make sure you involve him from the beginning. (Page 13).

Appendix 5 *(From page)*

Questionnaire: Project Integrity: Make sure the project has the right footing.

Questions	Yes	No	Answer/Comments
Do you know about Knowledge Management?			
What is the problem you/they are trying to solve?			
Why do you/they think is a problem?			
What is your definition of the words: Problem, Symptom, Solution, Tools,?			
What is the solution you envisioned?			
Is technology a solution or a tool?			
What is your definition?			
What are the goals of your project?			
What is the corporate vision defined by your leadership?			
Is there a correlation between that vision and the goals of your project?			
What are the success metrics for your project?			
How do you plan to bring users in, to share their knowledge?			
What is your policy against project failures?			

Appendix 6

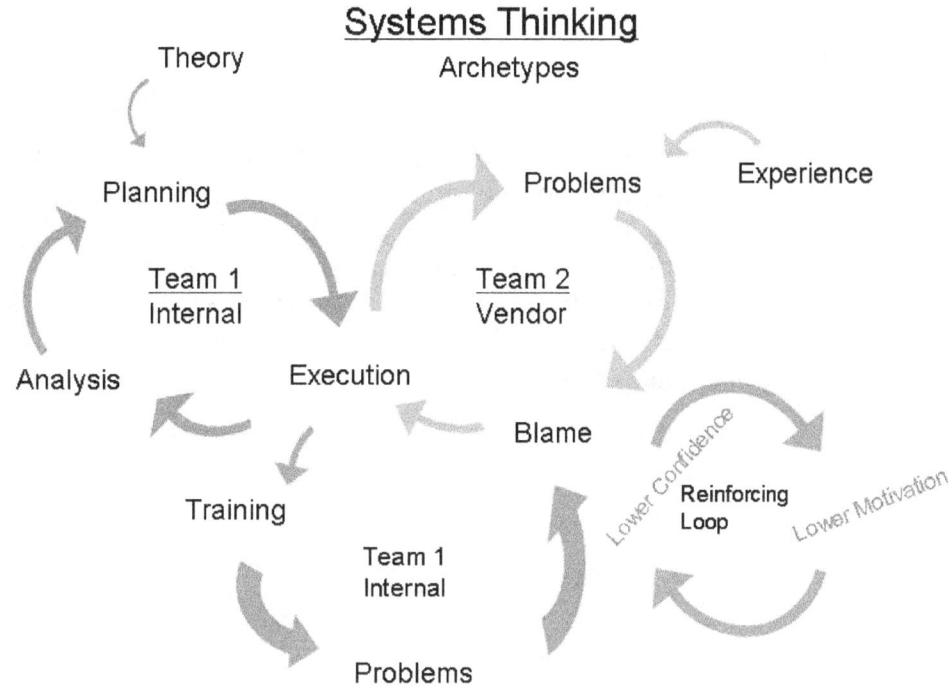

Figure 25 Systems Archetype

"Systems Thinking [21]is, more than anything else, a mindset for understanding how things work. It is a perspective for going beyond events, to looking for patterns of behavior, to seeking underlying systemic interrelationships which are responsible for the patterns of behavior and the events. Systems Thinking embodies a world-view. A world-view which implies that the foundation for understanding lies in interpreting interrelationships within systems. Interrelationships which are responsible for the manner in which systems operate. Interrelationships which result in the patterns of behavior and events we perceive.
Descartes and Bacon provided us with an analytic framework for understanding, and the scientific method. Newton, with the discovery of the laws of motion and gravity, provided us with a clockwork paradigm for understanding the universe. A paradigm which is not so much wrong as it is incomplete. The Newtonian paradigm embodies essentially a linear cause and effect relationship. A paradigm which is reinforced by the way in which we view daily events. The difficulty with this paradigm is that it

[21] http://www.systems-thinking.org/systhink/systhink.htm

provides a very limited short term perspective for understanding how things really work." By Gene Bellinger

In Figure 8, I analyzed what happens when a project has two teams ; The first is an internal team employed directly by the company, team two is an outsourced team, an external company providing technical expertise. The theory says that you can be successful having two or more teams in a project. My analysis shows to the contrary, that if one team is in charge of planning, and a second is in charge of executing, as soon they hit a problem, team two will very likely blame team one, because responsibility gets diluted if there is no authority that guides the teams as one (Whole). In this example, team one had to train users. So team two is dependent on team one, and team one is dependent on team two in the execution phase.

Appendix 7: How I found the meaning of Love.

Here is a little treasure I want to give to you. It might not have anything to do with the topic of this book, but it has a deep hidden connection, and I felt it was important to include it. I may someday write more about it. Yesterday, I realized that when Christ said "Do not resist evil", it meant that I needed not to resist it inside and outside because is part of Love.

I realized how obsessive I am, wanting to learn about the why of things. That obsession or passion depending how you see it, took me to high and lows, trying to find meaning and a guiding principle to my life. What I found in my knowledge journey was that we are all part of the same source, LOVE, but we are blind to the reasons why good things and bad things happen among many other things. I found they happen because is the way we can learn to LOVE GOD, LOVE OURSELVES, and LOVE OTHERS. Probably is the way to "human consciousness". That state of awareness and awakeness that permits human kind to transform itself from caterpillar, into a beautiful butterfly.

The conditions I needed to reach this conclusion were two: an open mind and bluntness to act. Through these, I learned to recognize valid information from dubious information and experience life in a somewhat different way (The hard way some might say). Acting (and resisting many times), allowed me to reflect and learn from my own fortunes and failures. In the midst of big challenges, that I now recognize where mostly in my own mind, and had nothing to do with the outside world, I finally learned.

The biggest lesson perhaps as seen through out this book and in my life is that knowledge without love, does not help us, only through love all things are possible for human kind, the beloved children of GOD.

Bibliography and references.

1- Tips for Effective Leadership Tip #5: Trust all the People all the Time by Wolf Rinke www.WolfRinke.com .
2- Backs to the future Aymara http://www.physorg.com/news69338070.html
3- Links embedded through out the book. This work is licensed under a Creative Commons Attribution 2.5 License.
4- The art of project management by Scott Berkun
5- The fifth discipline by Peter Senge.
6- Awakening Intuition by Mona Lisa Md Schulz
7- Guide to Project Management Body of knowledge by Project Management Institute.

Notes:

The outsourcing effect.

Outsourcing tasks, although seen beneficial in the short term, has bigger knowledge implications in the long term. The biggest might be that key knowledge about how to solve and most important, how to think successfully are retained by the consultant, and not the company. The snow ball effect I have seen is that a dependence is created so that if a given problem appears, instead of thinking through it, and imagining a solution within the company, managers bring consultants without a clearly defined process for knowledge transfer. But as we saw, you can only transfer information, but not the experience. So I could affirm that outsourcing really prevents you from learning unless you agree from the beginning with the consultants, that their role is to help you build a solution, but as importantly, teach your employees about how consultants did it in order to help employees solve their own problems. In reality, few consultants are willing to do it, because they want you to depend on them to solve the next batch of problems. A systemic problem that few understand, but that later will be more visible.

To contact the author, send an email to
leonardomorab@gmail.com